To Izen and Ingrid

Published by: Fontenelle Nature Association.
1111 N. Bellevue Blvd., Bellevue, NE 68005

ISBN 0-9748014-1-0

Front cover photograph by Neal Ratzlaff.
Back cover photograph by Tim Horning.
Front cover design by Tim Horning.
Interior book design by Tim Horning, and, Daniel MayBee of Omaha Print.
Printed by Omaha Print.

For additional copies of this book write to: Field Guide, Fontenelle Nature Association, 1111 N. Bellevue Blvd., Bellevue, NE 68005.

FIELD GUIDE TO

Trees, Shrubs, Woody Vines, Grasses, Sedges and Rushes

FONTENELLE FOREST & NEALE WOODS NATURE CENTERS

Neal S. Ratzlaff and Roland E. Barth

Fontenelle
Nature Association

ACKNOWLEDGMENTS

We thank **Gary W.** and **Marjorie M. Garabrandt** for helping us in many ways to produce this field guide, as well as the previous one. Marjorie's earlier survey of the vascular plants of the Fontenelle Forest and Neale Woods Nature Centers served as the baseline for this guide. Gary provided the background for most of the historical information outlined in the Introduction. We also thank other Fontenelle Nature Association staff members for contributing to this project, especially **Jason Krug, Erin McFadden, Mary Beth Steele and Laura Shiffermiller.**

We could not have produced a credible field guide without the expert advice and supervision provided by Professors Emeritus **David M. Sutherland** (University of Nebraska – Omaha) and **Robert B. Kaul** (University of Nebraska – Lincoln). We also thank **Steven B. Rolfsmeier** and **Barbara Wilson,** whose expertise and advice on the Nebraska sedges was invaluable. However, the responsibility for any errors in the identification or description of the species covered in this field guide rests solely with the two authors.

We thank **Chris Ratzlaff** for updating the trail maps used in both field guides. We also thank **Kristin Hoffman** for the drawings in the Illustrated Glossary depicting the features of grasses and sedges. We especially thank **Tim Horning**, who again played a major role in the design of this field guide, to include the cover, the title page and the amended Illustrated Glossary. **Laurine Blankenau** again contributed by reviewing and improving our text.

NEBRASKAland Magazine gave their gracious permission for the use of the first three pages of drawings in the Illustrated Glossary. These were created by **Tim Reigert**, Art Director of NEBRASKAland, for their *Field Guide to Wildflowers of Nebraska and the Great Plains*, by Jon Farrar.

We should also be thankful for the generous contributions of **Steve D. Hayes**, President of Omaha Print. Without his cooperation and support in underwriting a portion of the printing costs of this book, this project would not have been viable.

PREFACE

This field guide relates specifically to the Fontenelle Forest and Neale Woods Nature Centers. Its purpose is to enhance the experience of those who visit these two special places.

This book is designed to help visitors identify all those vascular plant species not addressed in our first book, *Field Guide to Wildflowers*. The first part describes the trees, shrubs and woody vines. The second part covers the sedges and rushes; the third part describes the grasses.

This guide also informs visitors where along the marked trails they might find a particular species, and it provides an estimate of its recent abundance. Together, these two field guides describe the considerable diversity of plant species which may be enjoyed and studied at our two Nature Centers.

Writing primarily for those who are just beginning to learn about plants, we tried to keep the descriptive terms used by botanists to a minimum. The botanical terms we did use are defined in the Glossary at the end of this book.

This field guide should also be of interest to the more experienced students of the flora. Most of the species described here will also be found in similar habitats well beyond our two Nature Centers.

Some of our oldest oak trees were already mature when the early explorers and settlers ventured up the Missouri River. Together with the many other native plant species still found in our two Nature Centers, they represent a part of our local ecosystems as they existed hundreds of years ago.

So we owe a great debt of gratitude to those who had the foresight to set aside these two natural areas, and to the many who contributed to their protection through the years. The considerable diversity of the flora still present and described in these two field guides serves as a testament to their efforts and as a challenge for future generations of stewards, volunteers and members.

Neal S. Ratzlaff

Roland E. Barth

CONTENTS

INTRODUCTION

Brief overviews of the Fontenelle Forest and Neale Woods Nature Centers are preceded here by a few introductory remarks about the local ecology and land management policies. More details about these two nature centers may be found on the FNA website at **www.fontenelleforest.org**.

Lay of the Land

Our local area is near the eastern edge of the Great Plains, which consisted primarily of grasslands in pre-settlement times. However, we are also located on the western edge of the so-called Midwest Oak Ecosystem.

In simplified terms, both nature centers had tallgrass prairies to their west. Oak savanna — grassland with scattered oak trees — covered much of the uplands. From there it transitioned to thicker oak woodlands and mixed forests closer to the bluffs overlooking the Missouri River. The bottomlands had willows, cottonwoods, sycamores and other plants more tolerant to periodic flooding.

Fires and floods helped to regulate this ecosystem. Periodic fires kept the prairies healthy and prevented other trees and woody plants from taking hold on the oak savannas. Spring flooding shaped the floodplains and defined their plant communities. From the prairies to the river's edge, this complex ecosystem sustained a considerable variety of native plant and animal species.

The relatively sparse populations of indigenous people seemed to have had little impact on the balance of this ecosystem. While they did cultivate, they were primarily hunters and gatherers who lived in lodges built on the high ridges overlooking the Missouri River.

When European settlers plowed the prairies during the second half of the 19th Century, they also suppressed the fires which had helped sustain the oak savannas. When they cut most of the more valuable timber, oak regeneration was stopped by faster-growing shade tree species. Many native understory plants were replaced as well. European settlers also introduced many of their own crop and decorative plant species. As a result, both nature centers have very different plant compositions today compared to those present 150 years ago.

Land Management

A generation ago land management philosophy was based on letting nature take its course without interference from man. FNA policy reflected the accepted practices of that time; natural areas were not to be "managed." In recent years that philosophy has evolved to more active roles for conservation, preservation and management.

In 1997, the FNA adopted a Land and Resources Management Plan which included not only protecting, but also restoring, recreating and maintaining native plant and animal communities within our two nature centers.

Fontenelle Forest Nature Center

The Fontenelle Forest Nature Center dates back to 1920, when a group of professionals led by Dr. A. A. Tyler and Dr. H. Gifford, Sr. acquired and set aside the first 300 acres of land. Additional parcels of adjoining land were bought in subsequent years, bringing the total to approximately 1400 acres today. Two Learning Centers, extensive boardwalks and a blind have recently been constructed to enhance the infrastructure. The website has more information regarding this.

The uplands are actually a patchwork of woodland parcels, each with its own land use history. With some exceptions, most of these parcels had been heavily logged, and some were used later on as pasture for milk cows. Other portions were cultivated as late as 50 years ago. Fortunately, quite a few old trees were spared, including a number of bur oaks dating back to the 1730s.

This logged upland forest was quickly overgrown by a variety of fast-growing shade trees. This prevented the natural regeneration of the previous oak savanna ecosystem. Despite all that, a varied mix of plant species still thrives in the uplands today, especially in the moist hollows and ravines.

In 1975, a small area along Prairie Trail was cleared to maintain a view of the Missouri River. It was seeded with native prairie species, some of which persist today.

The floodplain has also changed with time, especially since the Missouri River was channelized in the late 1930s, resulting in less spring flooding. In addition, the beavers have transformed large tracts of bottomland over the years. As a result some of the trails had to be rerouted or closed periodically.

As part of a habitat management effort, the Great Marsh and Hidden Lake were recently dredged, and Hidden Lake was reconnected to the Missouri River. A good variety of wildflowers and other plants, both native and naturalized, still thrive on the floodplain, especially near water. The open habitat along the railroad tracks promotes other plant species, most of them weedy varieties. Looking for wildflowers along the tracks is not recommended.

Signs of habitation will be discovered by hiking the trails, not only in the form of concrete foundations and other man-made objects, but also by the flowers planted long ago which still persist there. At the base of Mill Hollow, cultivated day lilies may be seen where a lumber mill once stood. Siberian squill may be found at Jim Baldwin's old farmstead overlooking the railroad tracks at the intersection of Prairie and History Trails. Boy Scout Camp Gifford had numerous buildings and was in operation until 1946. Some of the concrete foundations may be seen along Stream Trail. This stream was excavated in 1927 to connect the Great Marsh to the camp for canoe travel.

Land management efforts include studying the effects of excluding deer from a woodland site along Stream Trail. In this larger "exclosure," several uncommon plant species can be seen thriving today because they are protected from browsing. Efforts to control the overpopulation of deer with an annual hunt have been underway since 1996. A number of invasive species such as garlic mustard (*Alliaria petiolata*) are being controlled to impede their spreading and crowding-out of other understory plants. And a major effort is underway to recreate open savanna habitat on selected upland sites.

Neale Woods Nature Center

Early settlers in the Neale Woods area included three English brothers, David, Steven and George Neale. In the 1850's, George and his wife kept a hotel in the thriving community of Rockport which once existed along Rock Creek in the vicinity of the pond on River Road. David and Steven likely cut wood for the saw mill, clearing land that eventually became the Neale farm. In 1971, over a century later, Neale Woods Nature Center was established when David Neale's daughter, Miss Edith Neale, gave the 120-acre family farm to the Fontenelle Forest Association. She wished that the land, the site of present day Neale, Columbine, Settlers and Paw Paw Trails, be preserved as a memorial to her family and other early Nebraska pioneers. Only 4 years later, Carl A. Jonas gave 45 acres of his land to the FFA in memory of his father, one of its founding members. This tract includes present day Jonas Valley, Nebraska Prairie and adjacent woodlands. The following year Carl Jonas died, and his will gave 15 more acres plus his home to the FFA. The remodeled home, now the Carl Jonas Interpretive Center, and the lands at NWNC were opened to public use in 1983. Jonas' will also left substantial funds. Portions of it were used to purchase two additional properties. The first, an important 1977 acquisition of the 47-acre Koley Tract, physically connected the Jonas and Neale Tracts. This property includes present day Koley Prairie, an adjacent brome field and the woodland portion of Jonas Trail. The second land purchase was a strip of upland woods descending to Rock Creek along the west boundary, now traversed by Raccoon Hollow, Owl, Maidenhair and Hilltop Trails.

In 1995, a grant from the Lozier Foundation funded purchase of the 262-acre Krimlofski Tract, almost doubling the size of NWNC. It Included a sizable tract of Missouri River floodplain, a landform previously unrepresented at NWNC, and connected the uplands with the Missouri River. Portions of the buried old townsite of Rockport also lie within its boundaries. With the addition of two other small tracts, NWNC now encompasses a total of 562 acres including 8 miles of hiking trails.

Neale Woods, like Fontenelle Forest, has a variety of habitats and a more complex land use history than its present condition would suggest. Unfortunately, there is no survey similar to the remarkably detailed analysis of land use patterns at Fontenelle Forest made by Gary Garabrandt, FNA Director of Science and Stewardship. Therefore, conclusions about NWNC land use are more speculative. Aerial survey maps done in 1940 do show the entire Jonas Valley from Edith Marie Avenue to Nebraska Prairie under cultivation. Extensive portions of the Neale Tract were also cultivated fields, including most of the area south and east of present day Neale Trail and a broad upland ridge along upper Settlers Trail to its junction with Paw Paw Trail. These upland sites are still recognizable as young woodlands with few mature trees and sparse woodland wildflowers and grasses. Uplands with more mature woodlands and large, old bur oaks are confined mainly to northern portions of Hilltop Trail, the bluff overlooking Rock Creek next to lower Neale Trail, and the Missouri River bluff along History Trail. Spring woodland wildflower, native sedge and woodland grass displays are best in the older woods along the north facing slopes which descend to Rock Creek. These sites are accessed by Maidenhair, Paw Paw and Pond Trails. Beautiful older woodlands with good displays of woodland wildflowers, sedges and grasses also exist along lower Settlers and Columbine Trails and, closer to the Nature Center, along

Bittersweet and woodland portions of Jonas Trail. Much of the upland Krimlofski tract, the site of History Trail, was cultivated at the time of the 1940 aerial survey. The view today is great, and there is a good grass and sedge display at the overlook, but not many wildflowers. The floodplain beyond the bridge on Missouri River Ecology Trail was a part of the river bed at that time, so there is not much plant diversity, but a number of species preferring wetter habitats do provide for interesting viewing. Good displays also occur between the parking lot and the MRE Trail origin near the bridge over Rock Creek.

Recent additions to the NWNC flora include about 30 acres of restored prairies. Prairies occupied most of eastern Nebraska, but we really don't know how much of NWNC was prairie, so the word restoration may be presumptuous. Many prairie species were likely present, if only in open savannas along the ridges where they intermingled with spreading bur oaks. In any event the open fields of brome grass at newly opened NWNC soon became the target of prairie restoration efforts. This project has spanned some 20 years and provided new opportunities for education, interpretation, prairie grass and wildflower viewing. Upper and lower portions of Knull Prairie were seeded in 1983 and 85 with material gathered from local native prairies. Jonas and Koley Prairies followed in 1988. A Nebraska Environmental Trust grant enabled purchase of a commercial seed mix used for the 2001 Nebraska Prairie restoration. Another 2001 addition to the NWNC flora was the OPPD mitigation wetland in Jonas Valley. The Millard Transplant and a portion of the prairie in front of the Carl Jonas Interpretive Center originated from small bits of native Nebraska sod, transplanted from a site near 150th and Dodge Streets. Like all restorations, ours are not perfect representations of native prairies. They are considerably less diverse, and each contains introduced species that are out of their usual range. For fully a third of the approximately 100 prairie plant species in the NWNC restorations, fewer than 5 individual plants were found. Nonetheless, the common flowering plants and grasses native to our local prairies are quite well represented, and will reward the visitor with a good prairie experience.

Comments about NWNC would not be complete without discussing management of threats to our native woodlands and prairie restorations. Major threats to woodlands are from deer and exotic plants, most notably garlic mustard. Increasing evidence of damage to woodland plants from browsing deer prompted the first managed deer hunt at NWNC in 2003. All attempts to control the relentless advance of garlic mustard from its original foothold in the northwest portion of NWNC have been unsuccessful to date. Significant infestations now extend east into Settlers Hollow and south and east almost to the doors of the Carl Jonas Interpretive Center itself. The prairie restorations, of course, require fire management to control encroachment by woody vegetation. Burning each prairie every 4 years, supplemented by manual removal when necessary, has been reasonably effective.

What does the future hold for our prairies and woodlands? Most assuredly, they will be different from the 2006 "snapshot" provided by this book. Although some of the "players" may change, the protection afforded this special place by the Fontenelle Nature Association should provide visitors with a constantly changing, diverse seasonal plant display for many years to come.

HOW TO USE THIS FIELD GUIDE

Compared to the wildflowers covered in our first book, the plant families described in this field guide will be a bit more difficult to identify. In our closed, mature forests, binoculars may be required to study the leaves, flowers and fruits of the taller trees. At the other end of the scale, the subtle differences in some grasses and sedges will require a hand lens. So you may consider our first field guide to be a primer to the more difficult-to-identify species in this book.

Fortunately we only have a limited number of species to describe. That helps to make our job and that of our readers a lot easier. But it will take careful observation to identify some of the very similar species. In a few cases, the differences between closely related species are so minute and technical that we chose to treat them together, i.e. to lump them on one page.

The occurrence and abundance of flora seen along the trails of our two Nature Centers are constantly changing. A variety of factors play a role. The occurrence and abundance of annual species can vary greatly from year to year. Forests and other habitats continue to mature, thereby crowding out some of the understory species. Invasive species establish themselves, and efforts to control them may vary in effectiveness. From time to time some trails are rerouted or renamed, changing the habitats available for viewing. Therefore, the abundance we provide for each species represents but an average taken over the past several years.

Scope
This field guide describes all the vascular plants not covered in our first book. It has three parts: the trees, shrubs and woody vines; the sedges and rushes; and the grasses. Each part is prefaced with an introductory page.

We only describe those species found in areas and along trails currently accessible to the public. Therefore Camp Brewster is not included. We also do not include the cultivated species planted in areas adjoining the Visitors Centers, the Wetlands Learning Center, and around their parking lots.

Species Descriptions
For the most part, we describe one species per page. Several images representing typical specimens are usually sufficient to complement the descriptive text. For many of the trees, two facing pages are used, in order to show more of their features at different times of the year.

Names: The common name of each species is shown in large, bold letters. The scientific name follows, consistent with the recently published *The Flora of Nebraska* (see References). This scientific name consists of two parts. The first part is the genus; it is always capitalized. The second part is called the specific epithet. Together, these two names define a unique species. Both are italicized in this field guide, as are the scientific family names which follow. Sometimes species may be further subdivided into subspecies (ssp.) or varieties (var.). Then a third name is added after "ssp." or "var." An "x" indicates a hybrid.

Other Common Names: This heading is used when other common names are widely used for this species. These other common names are then included in the Index as well.

Flowering Period: An approximate flowering period is provided for each species. This time span is variable and may depend on specific location, annual weather variations and other reasons. We found that some plants bloom somewhat earlier in Fontenelle Forest than in Neale Woods.

Occurrence: The generally preferred habitat of each species is described first. When a habitat is unique to a species, it is underlined. Then we address their occurrence separately at Fontenelle Forest and at Neale Woods as appropriate. The names of the trails, prairies and structures are shown on the maps on the inside covers of this book. They are consistent with those in our first book. But be aware that a few of the trails have been rerouted or renamed since then. We use the following abbreviations:

FF ....Fontenelle Forest Nature Center
NW................Neale Woods Nature Center
BFLC............Katherine and Fred Buffett Forest Learning Center (FF)
WLC.............Gilbert M. and Martha H. Hitchcock Wetlands Learning Center(FF)
CJIC ....Carl Jonas Interpretive Center (NW)
MRE ....Missouri River Ecology (Trail) (NW)
GM ....Gifford Memorial (Boardwalk) (FF)

For each Nature Center, we provide an estimate of the abundance of a species, i.e.the chance of seeing one in a prairie or along a specific trail or two. The subjective definitions we use are provided here as a guide:

Abundant ....Easy to find; they're everywhere.
Common................Not hard to find.
Locally common ....Not hard to find at one or two locations.
Uncommon ....A few along that trail or area (4-10).
Rare........................Very few along that trail or area (1-3).

Descriptions: The format used to describe the various species depends somewhat on the type of plant considered. The trees and shrubs obviously require a different approach than the sedges and grasses. But we generally categorize plants as "native" or "introduced"; for the grasses and sedges we also categorize them as either "annual" or "perennial" (see the Glossary for definitions). Then the important visible features are described, with the most important diagnostic features underlined.

Identification: Whenever a more detailed comparison with similar species is warranted, a separate paragraph with this heading is used.

Comments: This heading is used when we had space for other information which may be of interest to our readers. We occasionally include information about the use of certain plants by Native Americans and the early settlers. However, we **do not** endorse any culinary or medicinal uses of any of these plants.

Trees, Shrubs and Woody Vines

This first part describes the largest and most visible plant families found in our two Nature Centers. For the purposes of this book, trees usually have a single thick trunk. Shrubs are smaller and they usually have multiple stems. Woody vines retain living stems during winter; they do not die back as do the herbaceous vines covered in *Field Guide to Wildflowers*.

The trees are described in this first section. There are 35 native and introduced species; each is shown on one or two pages. In addition, a few one-of-a-kind, probably planted trees are included at the end of this section, two per page. Those trees which had been planted near the Visitors Centers, and a few found far from the marked trails, were not included. To facilitate identification, the trees are arranged in sub-sections by the shape of their leaves. These sub-section numbers are included in the banner at the top of each of the species pages, e.g. TREES - 1:

        1 - Simple leaves with lobes
        2 - Simple leaves with or without teeth
        3 - Compound leaves
        4 - Palmate leaves and leaves as scales

The shrubs are described in the next section, with 24 native and introduced species, one per page. They are arranged roughly by the beginning of their flowering period, except for similar species, which are shown for comparison on facing pages. Additionally, a couple of one-of-a-kind shrubs, most likely planted, are depicted two per page. Shrubs planted next to the Visitors Centers were not included.

The woody vines complete this first part, with 10 native and introduced species, one per page. They are also arranged roughly by the beginning of their flowering periods, except when shown on facing pages to help in comparing with a similar species.

SILVER MAPLE
Acer saccharinum
MAPLE FAMILY *(Aceraceae)*

Flowering Period: Feb – Mar.

Occurrence: Floodplains. **FF:** Common along Hidden Lake Trail. **NW:** Uncommon along River Trail.

Description: This tall native tree has a trunk diameter of up to 3'. The mature bark has long, thin plates. This is the first plant to flower in our area with red or yellow male and female flowers on the same tree. The image at right top shows male flowers (left with drooping stamens) and a female flower (right). The leaves have silvery undersides. They are 3-5" long with 5 deeply cut, pointed lobes. The winged fruit (samara) is arranged in clusters of 2" long, paired "keys." They develop before the leaves do.

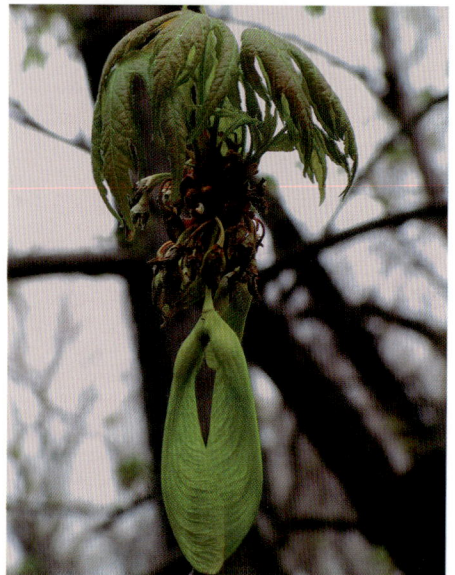

Identification: Box elder *(Acer negundo)* has similar samaras but very different leaves and flowers. Sugar maple *(Acer saccharum)* has leaves not as deeply lobed, and their undersides are not silvery.

DOWNY HAWTHORN
Crataegus mollis
ROSE FAMILY *(Rosaceae)*

Flowering Period: Apr – May.

Occurrence: Open wooded hillsides and floodplains. **FF:** Rare; look for one along the GM boardwalk. **NW:** Rare; single tree off Jonas Trail.

Description: This small native tree grows up to 25' tall. It has a rough and somewhat flaky bark. Its branches and twigs are <u>armed with stout thorns</u>. Winter twigs have bright red terminal buds. The <u>leaves and flowers usually appear together as "bouquets" at the ends of upright twigs</u>. The 3-5" leaves are variable in shape, lobed and sharply toothed. The white flowers have 5 sepals. The <u>fruit ripens to a red berry</u> in August; it is about ½" in diameter and resembles a rose hip.

Comments: Saplings of this declining tree species, along with bur oak *(Quercus macrocarpa),* have recently been planted on cleared ridges at FF as part of the oak savanna restoration project.

3

BUR OAK
Quercus macrocarpa
OAK FAMILY *(Fagaceae)*

Flowering Period: Apr – May.

Occurrence: Upland woods. **FF:** Common along Oak Trail. **NW:** Common along History Trail.

Description: This large native tree grows up to 80' tall and 3' plus in diameter. Its trunk and branches have deeply furrowed, gray bark. The leaves have rounded lobes. Winter twigs are light brown and fuzzy. The male flowers come in dangling catkins. Female flowers are small and inconspicuous on the same twig. The fruit, an acorn, grows solitarily or in clusters; it has fringes around its cup.

Identification: Red oak *(Quercus rubra)* has different bark, pointed lobes, and its acorns lack fringes.

Comments: The oldest tree in FF is a bur oak. Gary Garabrandt took core samples of the older trees in FF and tagged them near their base. No.263 was found to be "born" between 1722 and 1727. It's just off Oak Trail. Try to find it and pay your respects!

4

RED OAK
Quercus rubra; also
Quercus borealis var. maxima
OAK FAMILY *(Fagaceae)*

Flowering Period: Apr – May.

Occurrence: FF: Common in Childs Hollow.
NW: Common along wooded section of
Jonas Trail.

Description: This large native tree grows up
to 100' tall with a straight trunk up to 2' plus
in diameter. The mature bark has flattened
ridges. Winter twigs are brown and have a
terminal cluster of pointed buds. The mature
leaves are up to 9" long, with uneven,
pointed, bristle-tipped lobes. The younger
trees have striking green to red foliage in the
fall. The male flowers form dangling catkins;
the female flowers are on the same twig, but
small and inconspicuous. The fruit, an acorn,
matures singly or in pairs in the autumn of
the second year. A shallow cup holds the
ovoid nut.

Identification: Bur oak *(Quercus
macrocarpa)*, our other oak, grows mainly
on upland ridges. It has a bark with deeply
furrowed ridges, leaves with rounded lobes,
and acorn cups with fringes.

BOX ELDER
Acer negundo
MAPLE FAMILY *(Aceraceae)*

Flowering Period: Mar – Apr.

Occurrence: Woodlands, especially near water. **FF:** Common along Marsh Trail. **NW:** Common along MRE Trail.

Description: This medium-sized, native tree has a short trunk, up to 2' in diameter, often sprouting young branches as shown at right. The young branches are green and smooth. The compound leaves have 3 coarsely toothed or lobed leaflets. The reddish-brown terminal twigs are covered with a powdery white "bloom." Male and female flowers are found on separate trees. The male flowers (top row opposite) first appear reddish, later as dangling, green fascicles. The female flowers develop into paired keys (samaras), up to 2" long, either reddish or green (bottom row opposite).

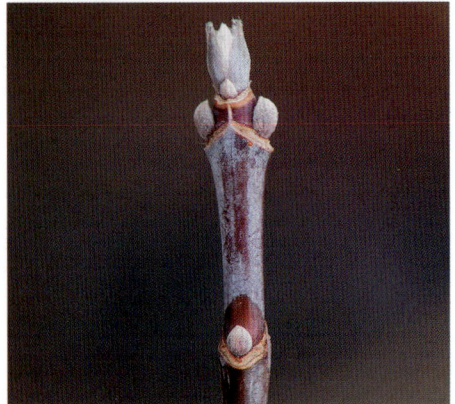

SYCAMORE
Platanus occidentalis
SYCAMORE FAMILY *(Platanaceae)*

Flowering Period: Apr – May.

Occurrence: Floodplains; at the northern edge of its natural range. When Prince Maximilian floated back down the Missouri River in 1834, he documented the first sycamore where the Boyer River joined, just northeast of Neale Woods. **FF:** Common along Walking Club Trail. **NW:** Common along MRE Trail.

Description: This massive tree grows up to 150' tall and 3' plus in diameter. Its bark varies from a "camouflage" pattern on young trees and branches, to a flaky brown and green closer to the base of older trees. The large leaves, usually with 5 pointed lobes, reach up to 10" across.

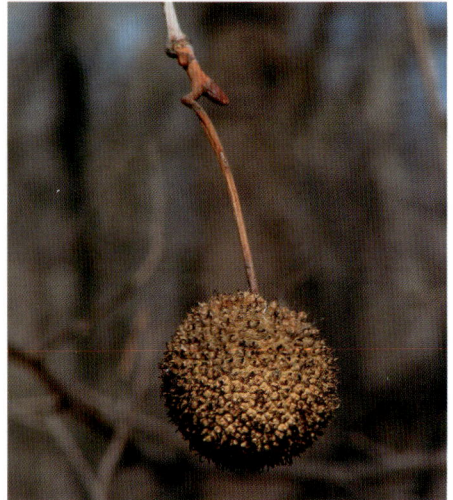

Miniature versions of the leaves (stipules) surround the leaf stalk. The seeds are tightly packed into light brown balls which dangle from stalks. These may be seen on some trees well into winter (opposite lower right).

AMERICAN ELM
(Ulmus americana)
ELM FAMILY *(Ulmaceae)*

Flowering period: Mar – Apr.

Occurrence: Along streams and in woodlands. **FF:** Common; look along the edge of the field at Camp Logan Fontenelle. **NW:** Common in Raccoon Hollow.

Description: This medium-sized native tree has light gray bark with deep, forking ridges. The leaves are alternate on the twigs, elliptic, sharply pointed, and usually unequal at their base. The leaf margins are doubly toothed, with small teeth along the lower side of the larger ones. A pair of lance-shaped stipules is found at the base of fresh leaves. The dark brown flower buds have smooth scales. The flower clusters, with unequal flower stalks, appear before the leaves. The fruit, a flat, elliptic samara, has a notched tip and a hairy margin; it appears in clusters by April.

Identification: We have three other elms. Siberian elm *(U. pumila)* has spherical buds and round samaras with smooth margins. Slippery elm *(U. rubra)* has buds with dense, red-brown hairs. Cork elm *(U. thomasii)* has slim buds and "corky" branches.

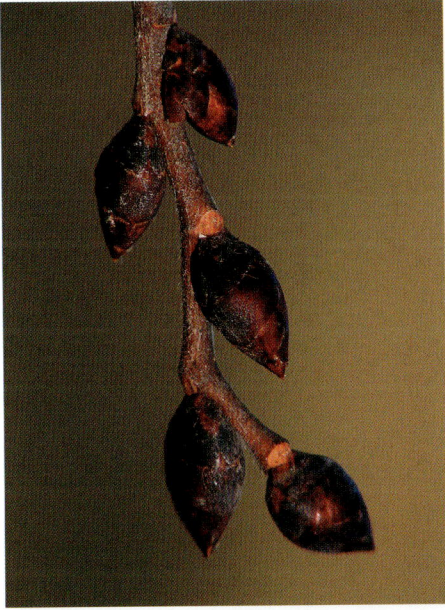

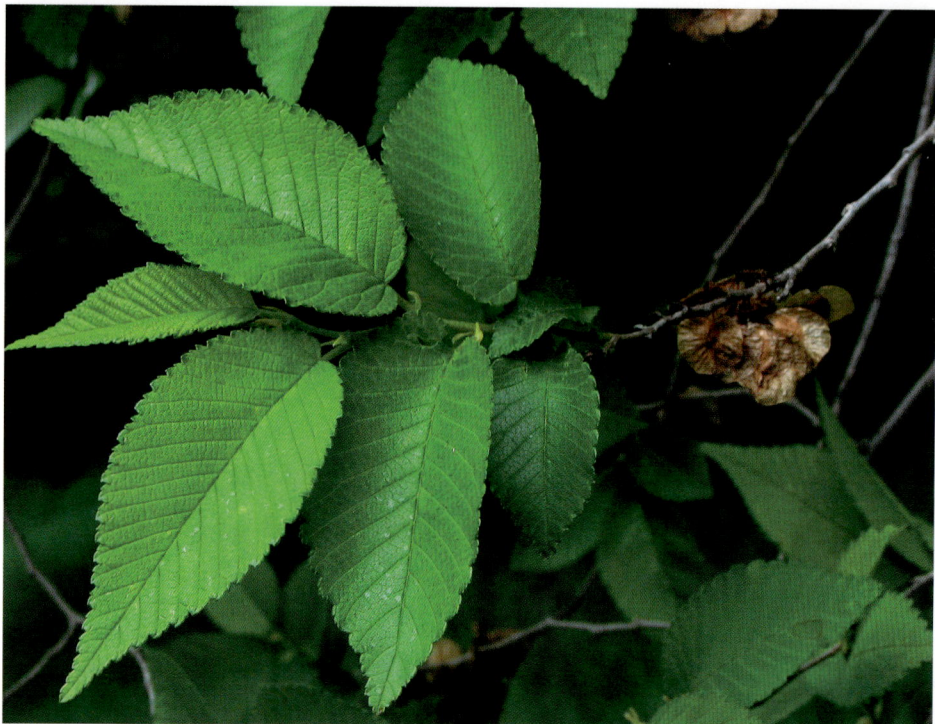

SIBERIAN ELM
Ulmus pumila
ELM FAMILY *(Ulmaceae)*

Flowering Period: Mar – Apr.

Occurrence: Where cultivated, and persisting elsewhere. **FF:** Uncommon; several along the edge of the field at Camp Logan Fontenelle. **NW:** Uncommon on upper Neale Trail.

Description: This medium-sized tree, a native of East Asia, was introduced here in the 19th Century. Most were planted, but many have escaped. The <u>leaves of this elm are usually smaller</u> when compared to our other elms. The base of the toothed leaves is nearly equal. The <u>winter buds are spherical, with hairs only on the margins of the scales</u>. The flowers form tight clusters on the twigs. The fruit is a round samara with a smooth margin and a notch at the tip.

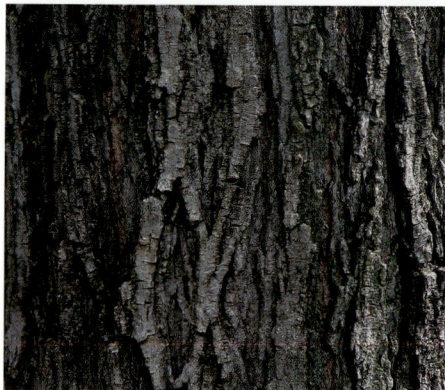

Identification: We have three other elms. American elm *(U. americana)* has larger leaves, pointed winter buds, and flowers and samaras in drooping clusters. The slippery elm *(U. rubra)* has larger leaves and fuzzy, red-brown winter buds. The cork elm *(U. thomasii)* often has "corky" lower branches, slim winter buds, and flowers and samaras with long stalks.

SLIPPERY ELM
Ulmus rubra
ELM FAMILY *(Ulmaceae)*

Other Common Name: Red elm.

Flowering Period: Mar – Apr.

Occurrence: Upland ravines and floodplains. **FF:** Uncommon; one overhangs Hidden Lake Trail about 50 yards east of the blind. **NW:** Uncommon in Raccoon Hollow.

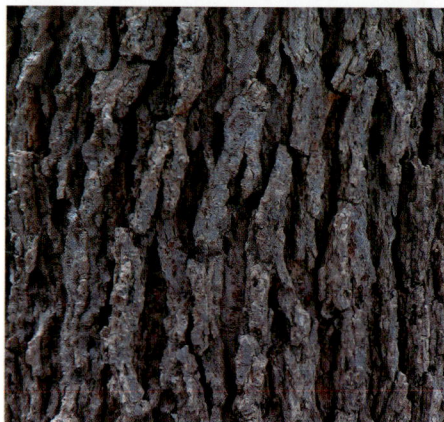

Description: This medium-sized native tree has gray bark with long, narrow ridges. The leaves are 4-8" long and rough to the touch. They are usually uneven at their base and end in a long, narrow point. The winter buds are found on hairy twigs; they are covered with dense, red-brown fuzz. The flower clusters, without stalks, turn into clusters of flat fruits (samaras), which have smooth margins and turn brown when mature.

Identification: We have three other species of elms. The American elm *(U. americana)* has smooth, pointed buds and elliptic samaras with hairy margins. The Siberian elm *(U. pumila)* has spherical buds, hairy only on the scale margins. Cork elm *(U. thomasii)* has slim buds, stalked flowers, and often "corky" younger branches.

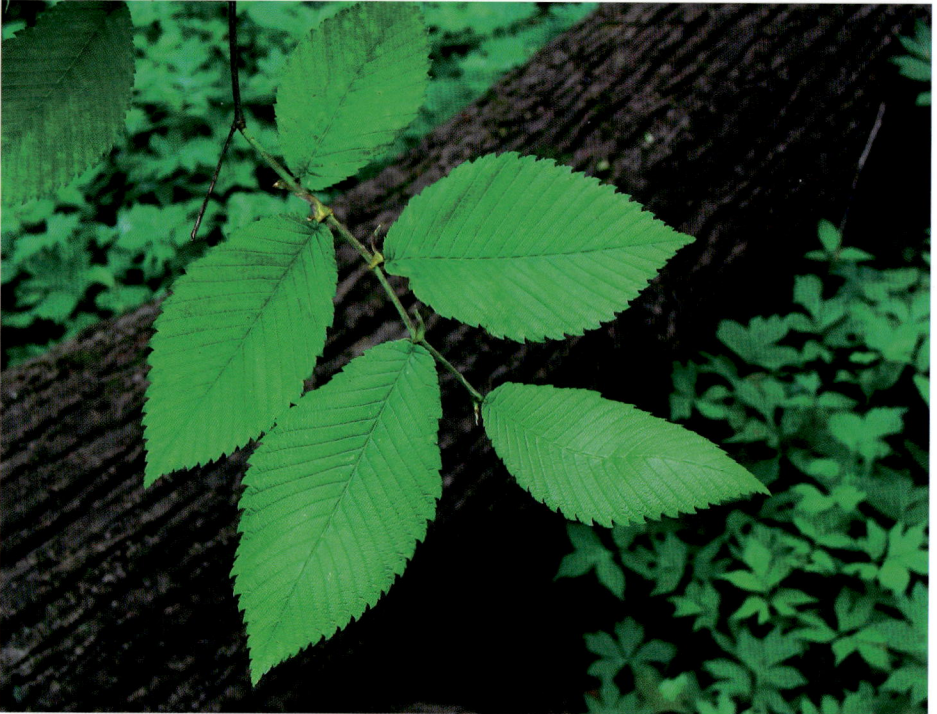

CORK ELM
Ulmus thomasii
ELM FAMILY *(Ulmaceae)*

Flowering Period: Mar – Apr.

Occurrence: Moist soils in valleys and along streams. **FF:** Uncommon in Childs Hollow. **NW:** Uncommon at the History Trail overlook.

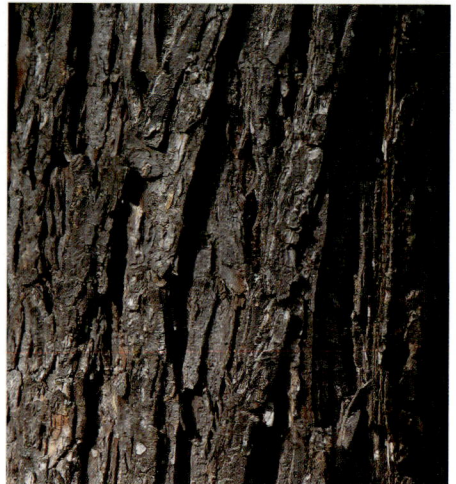

Description: This native grows as a small-to-medium-sized tree in our area. The bark is gray with thin, vertical ridges. The leaves, alternate on the twigs, are up to 6" long and 3" wide. The leaves are equal at the base. Lower branches often have unique "corky wings" (facing page). Pointed flower buds have appressed, silvery hairs. The flowers have long stalks and appear in drooping clusters before the leaves. The fruit is a thick, cup-shaped samara with dense hairs on its margins.

Identification: We have 3 other elms. The American elm *(U.americana)* has leaves often unequal at the base, and elliptic samaras. The Siberian elm *(U. pumila)* has spherical buds and smooth samara margins. The slippery elm *(U. rubra)* has buds covered with reddish-brown fuzz.

COTTONWOOD
Populus deltoides
WILLOW FAMILY *(Salicaceae)*

Flowering Period: Mar – Apr.

Occurrence: Mostly on floodplains. **FF:** Common along Walking Club Trail. **NW:** Common along MRE Trail.

Description: This massive native tree reaches heights of 150' and diameters of 6' plus. The bark of mature trees is gray and deeply furrowed. Dying trees often shed their thick bark, leaving behind tall, bare trunks. The winter twigs (opposite top row), often found on the ground after a storm, have sticky buds. Male and female flowers are on separate trees; they appear before the leaves. Both are borne on catkins, but female flowers are rarely seen from the ground. The reddish male catkins (middle row opposite) may be seen high in the tree or on the ground. The leaves are triangular,

sometimes wider than long. They turn bright yellow in the fall. The leaf stalks are flattened, allowing the leaves to flutter in the breeze to create that "murmur of the cottonwood trees." The seed capsules (opposite bottom left), found in drooping clusters, split into 2-4 parts to release vast numbers of small seeds attached to tufts of cottony hairs. They cover just about everything on the floodplains by June.

20

JUNE-BERRY
Amelanchier arborea
ROSE FAMILY *(Rosaceae)*

Other Common Names: Shad-berry;
shad-bush; service-berry.

Flowering Period: Mar – Apr.

Occurrence: Open woods, slopes and bluffs.
FF: Uncommon; several below the lowest
loop of Riverview Boardwalk. **NW:**
Uncommon along Settlers Trail above
Rock Creek.

Description: This small native tree has a
smooth bark and grows up to 25' tall. In
winter it shows red-brown twigs with buds
pointed upward. These trees are
conspicuous in late March and early April,
when they are among the first to bloom.
Dense white flower clusters appear well
before the oval, 2-4" long leaves. By June,
its green berries turn bright red. They are
usually eaten by birds before they ripen to
purple. By October these trees again become
conspicuous when their leaves turn a warm
reddish brown.

Comments: Native Americans mixed the
dried fruit with cornmeal to make bread.
Early settlers used the fruit in pies and
muffins.

IOWA CRAB
Pyrus ioensis
ROSE FAMILY *(Rosaceae)*

Other Common Names: Prairie crabapple;
bechtel crab.

Flowering Period: April.

Occurrence: We are on the western edge of
its natural range. **FF:** Rare; one tree was
found between the railroad tracks and
Walking Club Trail, about 25 paces west of
the intersection with Cottonwood Trail.

Description: This native tree, up to 12' tall,
has pink buds and white flowers which
bloom during the middle of April. Some of
its leaves are lobed and some of its twigs
have spurs. Its fruit, a crabapple, is less than
1" in diameter and ripens to greenish yellow
on 2" stalks by mid-August.

REDBUD
Cercis canadensis
CAESALPINIA FAMILY *(Caesalpiniaceae)*

Other common names: Eastern redbud, "Judas-tree".

Flowering Period: April.

Occurrence: In upland and floodplain woods. We are at the northern edge of its natural range. **FF:** Locally common on Redbud Trail. **NW:** Only a few planted trees near the CJIC and Neale Trail overlook.

Description: This native tree grows up to 20' tall, with a trunk up to 9" in diameter. The mature bark is scaly and gray. Pink, pea-like (legume) flowers appear in clusters well before its heart-shaped leaves. The fruit is a light green pod, about 3-4" long, which turns dark brown and often remains on the tree during winter.

PEACHLEAF WILLOW
Salix amygdaloides
WILLOW FAMILY *(Salicaceae)*

Flowering Period: Apr - May.

Occurrence: Along waterways and other wet areas. This is the most common willow in our area. **FF:** Uncommon along Marsh Trail. **NW:** Uncommon along MRE Trail.

Description: This native willow grows either upright to 50' or leaning over water. Its gray bark has deep, irregular furrows. Its winter twigs have large, reddish buds. The mature leaves, up to 6" long and 1" or more wide, are green on top and silvery below (see difference opposite). Male and female catkins are found on separate trees (male shown opposite above, female below).

Identification: Willows are difficult to identify. Our other native tree willow, the black willow *(S. nigra),* has much narrower leaves, green on both sides.

BLACK WILLOW
Salix nigra
WILLOW FAMILY *(Salicaceae)*

Flowering Period: Apr – May.

Occurrence: Along rivers and other moist places. **FF:** Rare; several may be seen along the Missouri River at Childs Hollow.

Description: This native erect or inclined tree has deeply furrowed bark and small, yellowish winter buds and twigs. The <u>mature leaves are long and narrow, less than 3/4" wide and green on both sides.</u> Male and female catkins are found on separate trees. Numerous yellow stamens make up the male catkins, <u>3-6 per individual flower (see upper image opposite)</u>. The female catkins produce seed capsules in May (opposite below).

Identification: Willows are difficult to identify. The peach-leaf willow *(S.amygdaloides)* has broader green leaves with silvery lower surfaces.

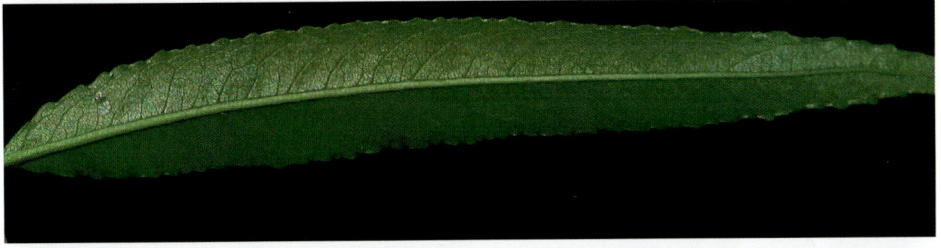

IRONWOOD
Ostrya virginiana
BIRCH FAMILY *(Betulaceae)*

Other Common Name: Hop-hornbeam.

Flowering Period: Apr – May.

Occurrence: Uplands. **FF:** Common along Hackberry Trail. **NW:** Common along Fox Trail.

Description: This small understory tree has a trunk up to 9" in diameter. The bark has distinctively long, narrow and scaly ridges. Its winter twigs (right) are hairy. The lance-shaped leaves are sharply toothed. The male catkins and a single female flower, about 1/2" long, are shown together in the upper left image on the opposite page. Its fruit is a papery cone, resembling hops. The dried, thin leaves may be seen on the tree throughout the winter, or until bundles of small male catkins appear at the ends of the twigs (opposite bottom).

RED MULBERRY
Morus rubra
MULBERRY FAMILY *(Moraceae)*

Flowering Period: Apr – May.

Occurrence: Moist floodplain woods. **FF:** Uncommon along Cottonwood Trail.

Description: This native medium-sized tree has brown or gray bark. The winter buds are greenish brown and pointed. The leaves are alternate on the twigs, up to 8" long and 5" wide, usually heart-shaped, but sometimes lobed when growing on vigorous branches. The lower leaf surfaces are rough and covered with short, fuzzy hairs. Male and female flowers are found on separate trees (male flowers opposite at left; female flowers to the right). The fruit is an edible, elongated berry, first red, then turning purplish black.

Identification: White mulberry *(Morus alba)* has smooth lower leaf surfaces.

WHITE MULBERRY
Morus alba
MULBERRY FAMILY *(Moraceae)*

Flowering Period: Apr – May.

Occurrence: Floodplains, especially near streams. **FF:** Common on Marsh Trail. **NW:** Common on MRE Trail.

Description: This medium-sized tree has a trunk which is often stained dark brown by oozing sap. The winter twigs have brown, spherical buds. The leaves are highly variable in size, shape, and texture. They are rough or shiny, often lobed, but the lower surfaces are always smooth. Male and female flowers are found on separate trees. The male flowers are shown opposite on the left; female flowers on the right. The fruit is a white, pink or nearly black berry.

Identification: Red mulberry *(Morus rubra)* has fuzzy hairs on lower leaf surfaces, and much less variable, mostly large, heart-shaped leaves.

Comments: This tree was introduced long ago from China as a favorite food for silkworms. But that venture failed.

HACKBERRY
Celtis occidentalis
ELM FAMILY *(Ulmaceae)*

Flowering Period: April.

Occurrence: Woodlands. **FF:** Common along Hackberry Trail. **NW:** Common along Hilltop Trail.

Description: This native tree grows large or small according to its habitat. Even small trees have the distinctively deep-furrowed gray bark. Alternate leaves are 2-4" long, lance-shaped, with sharp teeth. The inconspicuous flowers (right middle) appear with the leaves in April. The spherical fruit, about 3/8" in diameter, has a stalk up to 1" long. It is dark green by May and dark orange or purple by August.

Comments: Hackberry saplings quickly populate newly opened woodland as a dense shrub (as may be seen where woodland ridges had been cleared in FF to recreate oak savannas).

PAWPAW
Asimina triloba
CUSTARD APPLE FAMILY *(Annonaceae)*

Flowering Period: Apr – May.

Occurrence: Moist slopes and ravines. **FF:** Uncommon; look for a colony along Camp Gifford Road at the base of Signal Ridge Trail. **NW:** Uncommon along Paw Paw Trail.

Description: This native understory tree is found in colonies, propagating mainly by root sprouts. Winter twigs have pointed terminal buds. At maturity the leaves are 6-12" long, alternate on the stem, but they appear fan-shaped. The <u>striking flowers</u>, up to 2" across, have three <u>reddish purple petals;</u> they appear just before the leaves. The 3-5" long fruit, found singly or in clusters, ripens to yellow. Pawpaw rarely has fruit in our area.

Comments: The flesh of the ripe fruit tastes like banana custard. The leaves are host to the zebra swallowtail caterpillar *(Eutytides marcellus)*.

NORTHERN CATALPA
Catalpa speciosa
BIGNONIA FAMILY *(Bignoniaceae)*

Other Common Names: Cigar tree, Indian cigar tree, Indian bean.

Flowering Period: May – June.

Occurrence: Mostly on floodplains. **FF:** Uncommon; may be seen around the WLC parking lot. **NW:** Uncommon along MRE Trail.

Description: This non-native medium-sized tree, up to 90' tall, has red-brown, scaly bark. Winter twigs have large, oval leaf scars. The large leaves are variable in form, but mostly heart-shaped, up to 12" long and 8" wide. The large white flowers form upright, pyramid-shaped clusters (panicles). The fruit is a long bean-like pod, up to 18" long, which may be seen on some trees well into winter, long after its leaves have fallen.

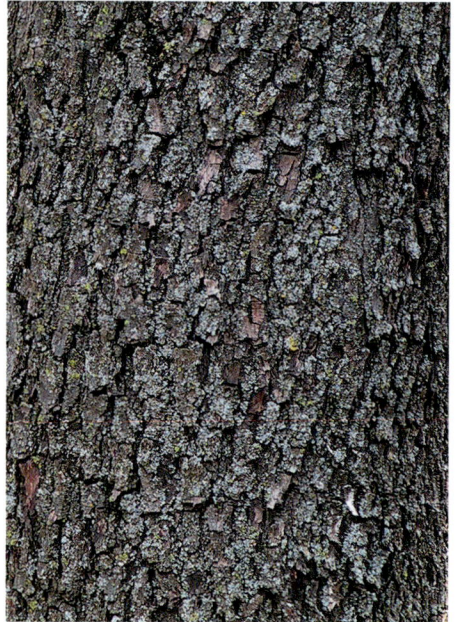

Comments: Extensively planted in our area as an ornamental tree and for fence post wood. It has widely escaped cultivation.

AMERICAN LINDEN
Tilia americana
LINDEN FAMILY *(Tiliaceae)*

Other Common Names: Basswood, bee-tree.

Flowering Period: Jun-July.

Occurrence: Mostly upland forests, but also in lowlands. **FF:** Common along upland trails. **NW:** Common along Paw Paw Trail.

Description: This large native tree grows to 120' tall, typically with two or more straight trunks emerging from the same old stump or root system. The young bark shows brown fissures (right); the older bark has narrow, scaly ridges. The dense foliage consists of large, heart-shaped leaves with stalks, alternate on the stems, darker green above than below. The winter buds are bright red. The yellow-white flowers are fragrant and come in long-stalked clusters, dangling from a unique, leafy bract. The fruit is spherical, less than ½" in diameter and covered with fuzz

Comments: Bees swarm the abundant flowers in season and produce a honey with a distinctive taste. Native Americans made ropes and mats from the inner bark of this tree.

RUSSIAN OLIVE
Elaeagnus angustifolia
OLEASTER FAMILY *(Elaeagnaceae)*

Flowering Period: May – Jun.

Occurrence: Where planted, and in open areas and woodland edges. **FF:** Rare at Camp Logan Fontenelle. **NW:** Uncommon along edge of Jonas Prairie next to the entrance road.

Description: This cultivated and naturalized tree grows up to 25' tall; it is recognized from afar by its silvery leaves, which are slender, up to 3" long and alternate on the twigs. The bell-shaped flowers with 4 pointed lobes are orange to yellow inside. Clusters of 1-3 flowers form at the leaf nodes. The berry-like fruit ripens in September; it is tan or silvery.

Identification: The closely related autumn olive *(Elaeagnus umbellata)*, a large shrub, has oval leaves, light yellow flowers and red fruit.

BLACK WALNUT
Juglans nigra
WALNUT FAMILY *(Juglandaceae)*

Flowering Period: May – Jun.

Occurrence: Mostly upland woods, but also on floodplains. **FF:** Uncommon in Childs Hollow. **NW:** Uncommon in Settlers Hollow.

Description: This large native tree has a trunk up to 4' in diameter, with a deeply furrowed, variable bark pattern (note different pattern on the same trunk shown at right).The winter twigs have rounded, fuzzy buds above three-lobed leaf scars. The compound leaves have 12-20 slender leaflets not quite opposite on the stalk. The spherical fruit has a smooth, pale yellow husk. It turns black as it matures, yielding a rough black nut up to 1½" in diameter.

Identification: Tree-of-Heaven *(Ailanthus altissima)* has similar leaves but different buds and fruit.

43

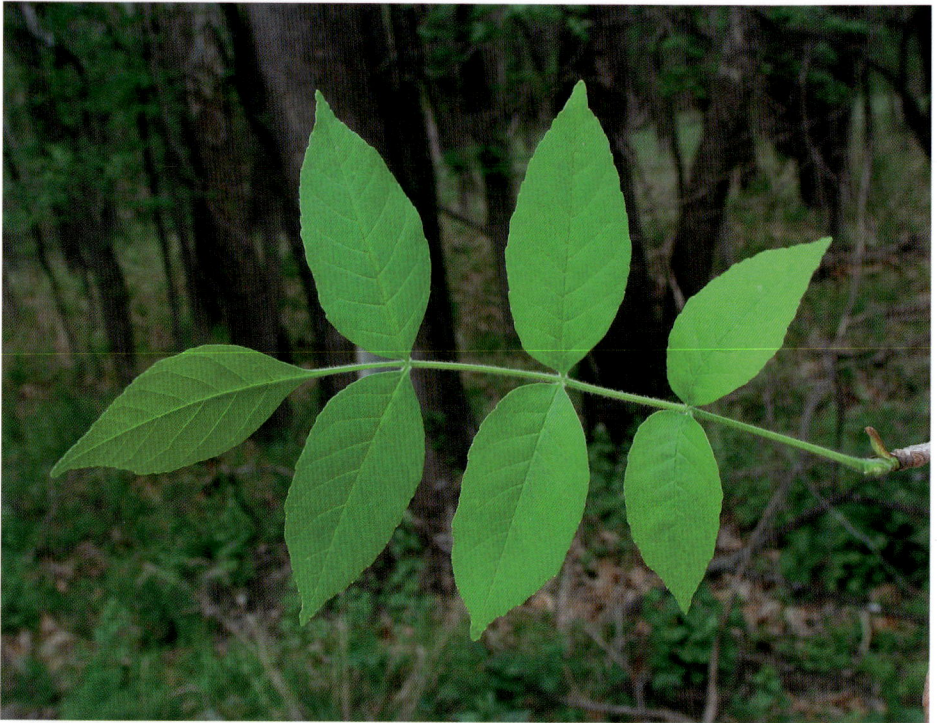

GREEN ASH
Fraxinus pennsylvanica
OLIVE FAMILY *(Oleaceae)*

Flowering Period: Apr – May.

Occurrence: Mostly on floodplains. **FF:** Common along Camp Gifford Road. **NW:** Common along Nebraska Trail.

Description: This native, medium-to-large tree grows up to 80' tall and has gray, deeply furrowed, diamond-shaped bark. The winter twigs have pointed terminal buds and semicircular leaf scars with a flat top (opposite top). Male and female flowers are found on separate trees (second row opposite, left and right respectively). The leaves are opposite, up to 12" long. They consist of 7 or 9 lance-shaped leaflets without (or with very short) stalks. The leaves turn a bright yellow in the fall. The fruit, a samara, appears in dense clusters, first green then light brown.

Identification: Similar to white ash *(Fraxinus americana)*, which has longer leaflet stalks; rounded terminal buds; crescent-shaped leaf scars; more compact samaras (compare typical examples on opposite page); and the leaves of many of the white ash trees turn purple in the fall.

Comments: This tree was often planted as a windbreak on prairies. Its wood was also used for tool handles and other implements, just as white ash *(Fraxinus americana)* was.

44

WHITE ASH
Fraxinus americana
OLIVE FAMILY *(Oleaceae)*

Flowering Period: Apr – May.

Occurrence: Mostly uplands and ravines. **FF:** Uncommon on Riverview Boardwalk. **NW:** Uncommon along Hilltop Trail.

Description: This native medium-to-large-sized tree has a straight trunk with gray, variable bark. The winter twigs have rounded terminal buds and crescent-shaped leaf scars (opposite top). The leaves are opposite and up to 12" long, consisting of 7 or 9 lance-shaped leaflets on stalks ¼" or longer. The leaves are whitish below; they turn yellow or purple in the fall. Male and female flowers (opposite second row left and right respectively) are on separate trees. The fruit, a samara appears in dense clusters by May; it turns brown by June.

Identification: Green ash *(Fraxinus pennsylvanica)* has leaflets paler below but not whitish, leaflet stalks less than 1/4", leaves turning only yellow in fall, and samaras with longer "wings" and a flatter seed (compare typical samaras on opposite pages).

Comments: Young ash trees were used by the early explorers and settlers to make and replace oars, axe handles and other hand tools. White ash is still used to make baseball bats.

SHAGBARK HICKORY
Carya ovata
WALNUT FAMILY *(Juglandaceae)*

Flowering Period: Apr – May.

Occurrence: Mostly in upland woods. **FF:** Common on Hickory Trail. **NW:** Uncommon along Hilltop Trail.

Description: This native tree grows up to 80' tall with a trunk up to 2' in diameter. The bark is smooth on young trees. On mature trees the bark is very shaggy (right). The compound leaves, up to 16" long, have 3 or 5 broad leaflets. The leaves turn yellow, then brown in the fall. The dried, brown leaves can often be seen on the tree well into winter. The fruit is a large, edible nut, which is enclosed in a thick, spherical husk with 4 grooves (opposite, lower left). The winter buds are large, tan with loose bud scales as shown opposite, lower right.

Identification: The bitternut hickory *(Carya cordiformis)*, our only other hickory, has bright yellow winter buds, smaller leaves with 7 or 9 leaflets, pointed and ridged husks; and its mature bark is not shaggy.

BITTERNUT HICKORY
Carya cordiformis
WALNUT FAMILY *(Juglandaceae)*

Flowering Period: Apr – May.

Occurrence: Mostly upland woods, especially hollows and ravines. **FF:** Common in Childs Hollow. **NW:** Uncommon in Raccoon Hollow.

Description: This medium-sized native tree has a smooth bark with vertical "stretch marks" on young trees and branches (near right). Mature trees have a furrowed bark (far right). Male and female flowers are found on the same tree, but they are seldom visible from the ground. Its solitary or paired fruit is a pointed husk with 4 ridges. It splits along those ridges when ripe to yield a bitter fruit.

The compound leaves, 9-12" long, have 7 or 9 lance-shaped leaflets. They turn bright yellow in the fall. The distinctive winter buds are bright yellow and lack scales.

Identification: Shagbark hickory *(Carya ovata)*, our other hickory, has large, brown winter buds, shaggy bark on mature trees, and larger leaves, usually with 5 broad leaflets.

KENTUCKY COFFEE-TREE
Gymnocladus dioica
CAESALPINIA FAMILY *(Caesalpiniaceae)*

Flowering Period: May – Jun.

Occurrence: Rich woods, along streams and open wooded hillsides. **FF:** Uncommon in Mill Hollow. **NW:** Uncommon in Raccoon Hollow.

Description: This native tree grows up to 100' tall. The <u>leaves are twice (bipinnately) compound, with oval, pointed leaflets</u>. Purple and white tube flowers form upright clusters (racemes). The <u>fruit is a thick, brown pod with a powdery, white coating (bloom)</u>. The fruit can often be seen on bare trees during winter. The gray bark has narrow, scaly ridges, often turned up on one side.

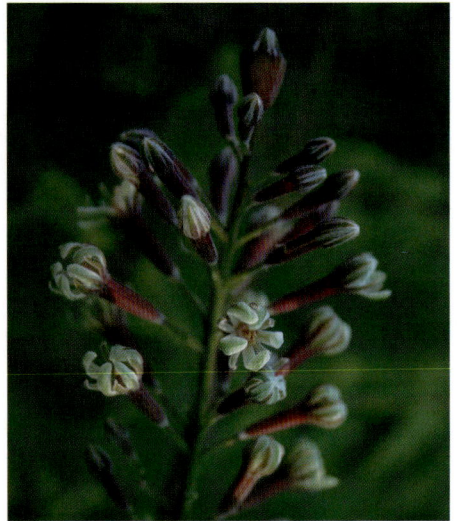

Identification: Honey locust *(Gleditsia triacanthos)* has similar-looking bark, but it has thorns, rounded leaflets, and elongated, flat seed pods.

HONEY LOCUST
Gleditsia triacanthos
CAESALPINIA FAMILY *(Caesalpiniaceae)*

Flowering Period: May – Jun.

Occurrence: Floodplains and uplands. **FF:** Uncommon along Cottonwood Trail. **NW:** Uncommon in Raccoon Hollow.

Description: This large, <u>thorny tree</u> grows up to 90' tall. Young trees have <u>red thorns</u> and a smooth bark. Mature trees have <u>brown thorns</u> and scaly bark ridges curving outward. It has twice (bipinnately) compound leaves with 14-20 <u>rounded leaflets</u>. The <u>fruit is a flat, 6-16" long, twisted seed pod</u>, first yellow, then dark brown, often remaining on the tree well into winter.

Identification: The related Kentucky coffee-tree *(Gymnocladus dioica)* has <u>similar bark but lacks thorns, has pointed leaflets, and thick seed pods.</u>

BLACK LOCUST
Robinia pseudoacacia
BEAN FAMILY *(Fabaceae)*

Flowering Period: May.

Occurrence: FF: Locally common at the old farmstead along Prairie Trail; planted long ago and propagating there mainly from root sprouts.

Description: This medium-sized tree has a rough, scaly bark. The branches have spines. The compound leaves have up to 21 rounded leaflets. The <u>white, fragrant legume flowers form drooping racemes</u>. The fruit is a pod with 3-8 brown, kidney-shaped seeds.

TREE-OF-HEAVEN
Ailanthus altissima
QUASSIA FAMILY *(Simaroubaceae)*

Flowering Period: May – Jun.

Occurrence: Floodplains and ravines. **FF:** Uncommon on waste ground, especially along railroad tracks. **NW:** Uncommon in Hilltop brome field.

Description: This medium-sized, invasive tree spreads rapidly, mainly by root suckers. Its straight trunk has shallow ridges. The large leaves have up to 41 lance-shaped leaflets opposite on the stem. Cut trees produce <u>numerous root suckers</u> with prominent leaf scars (opposite, top left); they grow rapidly (opposite bottom left and right). Male and female flowers are on separate trees. The small, greenish male flowers grow in upright clusters (opposite top right). <u>The winged fruits (samaras) form massive clusters</u>; they are first green or pinkish, later light brown.

Comments: This tree is considered a threat to our native species; it is controlled by cutting and by treating the stumps.

WESTERN BUCKEYE
Aesculus glabra
BUCKEYE FAMILY *(Hippocastanaceae)*

Other Common Name: Ohio buckeye.

Flowering Period: April.

Occurrence: Woodlands. **FF:** Rare; one tree found off Walking Club Trail, another along Signal Ridge Trail. A number of small saplings leaf out each year along Riverview Boardwalk. They are usually browsed by late summer.

Description: This small tree grows up to 40' tall with a diameter of 1-2'. The bark is scaly brown. Its palmate leaves emerge early in the season from large buds. The yellow flowers are arranged in tall panicles. The fruit is a spiny capsule, larger than 1" in diameter. It holds a brown chestnut. The two trees found in FF flowered, but did not bear fruit in 2006.

RED CEDAR
Juniperus virginiana
CYPRESS FAMILY *(Cupressaceae)*

Other common names: Eastern red cedar; red juniper.

Flowering period: Apr – May.

Occurrence: Planted, or growing freely in prairies, where they are controlled by cutting. **FF:** Rare; mostly stunted trees less than 3' tall. **NW:** Planted as a windbreak along the edge of Koley Prairie; rare elsewhere.

Description: This native evergreen grows to 60' tall, with male and female cones on separate trees. Male trees appear more yellowish green than the darker green female trees. Close-ups of the yellow male cones and light blue female berry-like cones are shown at right. The leaves are in the form of tiny, overlapping scales. The bottom image shows a common fungus, cedar-apple rust, which also infects apple trees.

SUGAR MAPLE
Acer saccharum
MAPLE FAMILY *(Aceraceae)*

Description: This solitary tree grows in FF near the kiosk on the Childs Hollow dike. It had been planted there in 1935 from seeds originating in Ohio. It displays distinctive red foliage in October. Another sugar maple, planted between the street and the FF Visitors Center parking lot, may be studied more readily, since the leaves and fruit hang closer to the ground. But this planted tree may be of another variety.

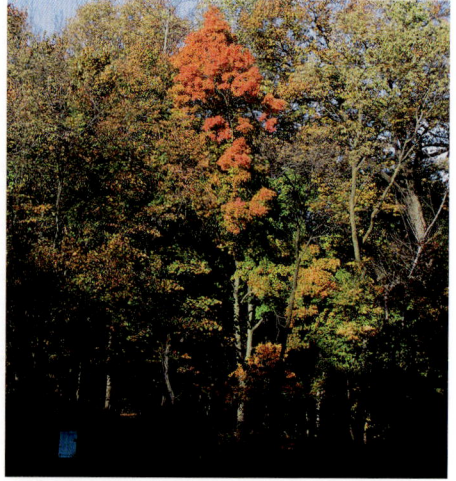

GOLDEN WEEPING WILLOW
Salix x sepulcralis var. chrysocoma
WILLOW FAMILY *(Salicaceae)*

Description: This tall, solitary tree grows in FF about 50 yards southeast of the Camp Gifford Road railroad crossing. It has strongly pendulous (weeping) branches. The golden branches can best be seen in March and April from Stream Trail, from about 25 paces south of the road, when all the other trees are still a drab gray. This tree was probably planted during the first half of the 20th Century. Willows come in male and female trees. This one is a male tree.

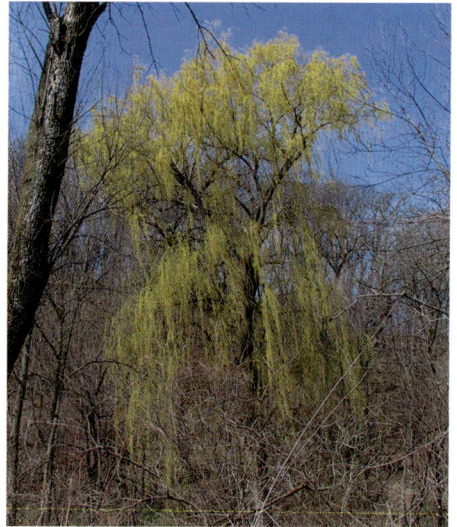

CRABAPPLE
Malus sp.
ROSE FAMILY *(Rosaceae)*

Description: This solitary tree, about 15' tall, was probably planted long ago along South Stream Trail. Its large pink flowers may be seen in April. The green apples, about an inch in diameter, dangle in bunches from 2" long stalks.

Comments: There are numerous ornamental varieties of crabapple. The casual observer will only notice this tree when it is in bloom. Later in the season it usually is draped by woody vines.

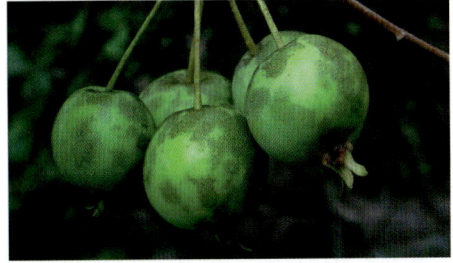

CRABAPPLE
Malus sp.
ROSE FAMILY *(Rosaceae)*

Description: This small tree is located on the stream side of South Stream Trail, not far north of the crabapple described above. Its dark red buds appear in April. They soon open up to form large, pink flowers. Its red fruit is about 1/2" in diameter; it dangles in bunches from 1 1/2" stalks.

Comments: This is another variety of ornamental crabapple. We are not sure why these two crabapples were planted at this particular location. We do know that the Boy Scouts planted some trees after the stream was excavated in 1927 to connect the Great Marsh with the Boy Scout camp for canoe travel (see Introduction). But these don't seem quite old enough.

HAZELNUT
Corylus americana
BIRCH FAMILY *(Betulaceae)*

Flowering Period: March.

Occurrence: Upland woods and thickets. **FF:** Uncommon near the entrance to Mormon Hollow. **NW:** There is a large colony on the north boundary along County Road P-40.

Description: This native shrub grows up to 12' tall. Male and female flowers are found on the same plant. Small <u>male catkins are present during winter</u>. They enlarge and flower along with a few <u>tiny red female flowers</u> by mid-March, well before the emergence of leaves, which are ovate, toothed and up to 5" long. The young twigs are hairy. The <u>fruit, a wrapped nut, develops singly, or as one or two clustered pairs</u> by late July.

Comments: The hazelnut, also known as filbert, is edible. The Omaha-Ponca collected these along with other nuts. They ate them raw, with honey, or in a soup.

PRICKLY ASH
Zanthoxylum americanum
CITRUS FAMILY *(Rutaceae)*

Other Common Name: Toothache tree.

Flowering Period: Early April.

Occurrence: Roadsides, near water. **FF:** Common along Stream Trail. **NW:** Rare along Jonas Trail.

Description: This native shrub forms thickets up to 12' tall. The light gray stems are armed with <u>stout spines</u>. Its compound leaves have 5-11 elliptic or ovate leaflets, about 3" long. <u>Male and female flower clusters are on separate plants</u>; they appear before the leaves. Female flowers are green with reddish tips (left). Male flowers (right) have prominent, yellow stamens. The <u>fruit is a rough red follicle which releases a shiny black seed</u> in August.

Comments: Bruised leaves have a citrus smell. Native Americans chewed the leaves and bark to relieve toothaches, hence the other common name.

WILD PLUM
Prunus americana
ROSE FAMILY *(Rosaceae)*

Flowering Period: Apr – May.

Occurrence: Along forest edges and open areas. **FF:** Rare and decreasing; several near the northern crest of Oak Trail. **NW:** Rare along the edge of Koley Prairie; a few on the northern boundary along County Road P-40.

Description: This native shrub or small tree usually spreads by root sprouts to form thickets. It is one of the first shrubs/trees to bloom. The winter twigs end in numerous smooth, pointed buds. The white flowers, with 5 petals, appear in tight clusters. The fruit is an oblong or globe-shaped plum, first green, then pink or reddish by August.

Other Comments: Highly valued and eaten fresh or dried by local Native Americans. The Omaha timed planting corn, beans and squash to the blooming of the wild plum.

ROUGH-LEAVED DOGWOOD
Cornus drummondii
DOGWOOD FAMILY *(Cornaceae)*

Flowering period: May – Jun.

Occurrence: Wet or dry woods. **FF:** Abundant along floodplain trails. **NW:** Abundant along the edges of prairies.

Description: This native shrub grows up to 20' tall. The lance-shaped leaves, up to 5" long and rough to the touch, are opposite on reddish brown, new-growth twigs. The white flowers, with 4 petals, form a dense, flat-topped cluster (cyme). The fruit is white but occasionally light blue and spherical.

Identification: Alternate-leaved dogwood *(C. alternifolia)* has smooth, alternate leaves and dark blue fruit. Pale dogwood *(C. amomum)* has smooth leaves, maroon-colored branches in winter and blue fruit.

Comments: This shrub is highly invasive in prairies. It is controlled there by cutting and periodic burning.

ALTERNATE-LEAVED DOGWOOD

Cornus alternifolia
DOGWOOD FAMILY *(Cornaceae)*

Flowering Period: Apr – May.

Occurrence: Moist ravines and woods. **FF:** Rare; several plants were found near the entrance to Handsome Hollow. Our area is outside their natural range; they were probably planted there long ago.

Description: This native shrub or small tree, up to 15' tall, has up to 4" diameter trunks. The oval <u>leaves are smooth,</u> with stalks of variable lengths. They are <u>alternate on their twigs but appear fan-shaped</u>. The white flowers form flat-topped clusters. The <u>fruit is dark blue and nearly spherical.</u>

Identification: Rough-leaved dogwood *(C. drummondii)* has rough, opposite leaves and mostly white fruit. Pale dogwood *(C. amomum)* has opposite leaves and maroon stems.

PALE DOGWOOD
Cornus amomum
DOGWOOD FAMILY *(Cornaceae)*

Flowering Period: May – Jun.

Occurrence: Moist woods, stream banks.
FF: Rare; one clump where Stream Trail crosses Camp Gifford Road; another at the base of Mill Hollow. Both were probably planted.

Description: This native shrub grows in clumps up to 15' tall. It has brown stems which turn maroon in winter. The leaves are lance-shaped, smooth to the touch, and opposite on green twigs. The white flowers have 4 petals and form flat-topped clusters (cymes). The fruit is a blue drupe.

Identification: Alternate-leaved dogwood *(C. alternifolia)* has leaves alternate on the twigs and dark blue fruit. Rough-leaved dogwood *(C. drummondii)* has rough leaves and spherical, mostly white fruit.

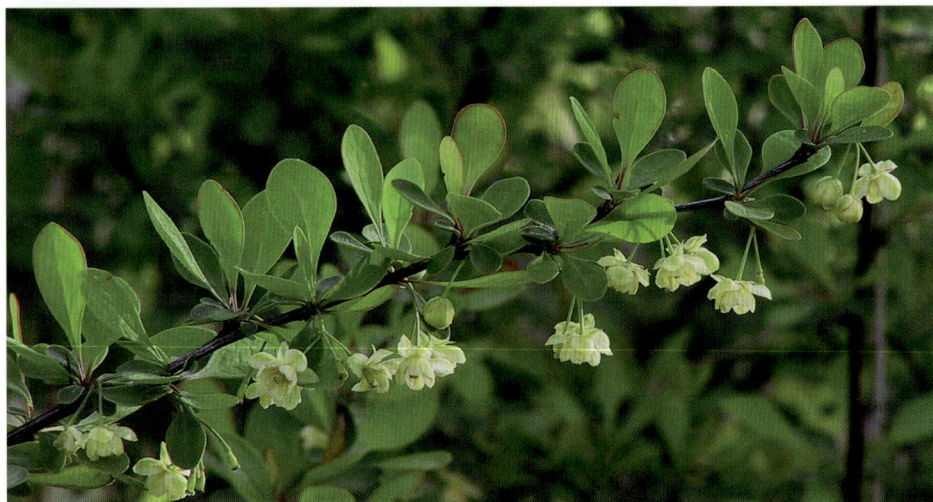

JAPANESE BARBERRY
Berberis thunbergii
BERBERIS FAMILY *(Berberidaceae)*

Flowering Period: Apr – May.

Occurrence: Mostly on floodplains, but in upland woods as well. **FF:** Uncommon along Cottonwood Trail. **NW:** Rare along Paw Paw Trail.

Description: This native of Japan was introduced to this continent as an ornamental shrub. It is considered a noxious weed today. It grows 3-6' tall and has spoon-shaped leaves in clusters alternate on straight twigs. These leaf clusters are joined by a bundle of cream-colored flowers on stalks (pedicels) and a single thorn. The leaves turn yellow, later red by October. The fruit is a bright red, egg-shaped berry, which can often be seen on this shrub well into winter.

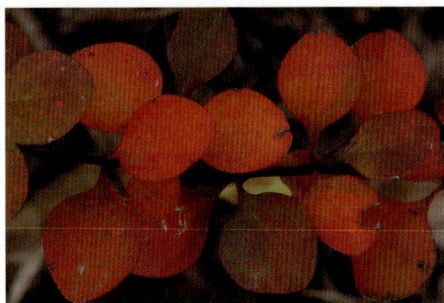

Comments: This highly invasive plant is being controlled by uprooting.

EASTERN CHOKE CHERRY
Prunus virginiana
ROSE FAMILY *(Rosaceae)*

Flowering Period: Apr – May.

Occurrence: Open areas, including roadsides and ravines. **FF:** Uncommon at base of Childs Hollow. **NW:** Uncommon at bluff's edge on History Trail.

Description: This native shrub or small tree grows up to 20' tall. It often forms thickets. The alternate leaves are oval and pointed, with fine teeth on the margins. New growth twigs and leaves are often reddish. The white flowers form crowded, upright clusters (racemes) at the ends of twigs. The fruit ripens by July as dark red-to-blackish berries, dangling on stalks like grapes.

Comments: Choke cherries are astringent to the taste, but make fine jellies and preserves.

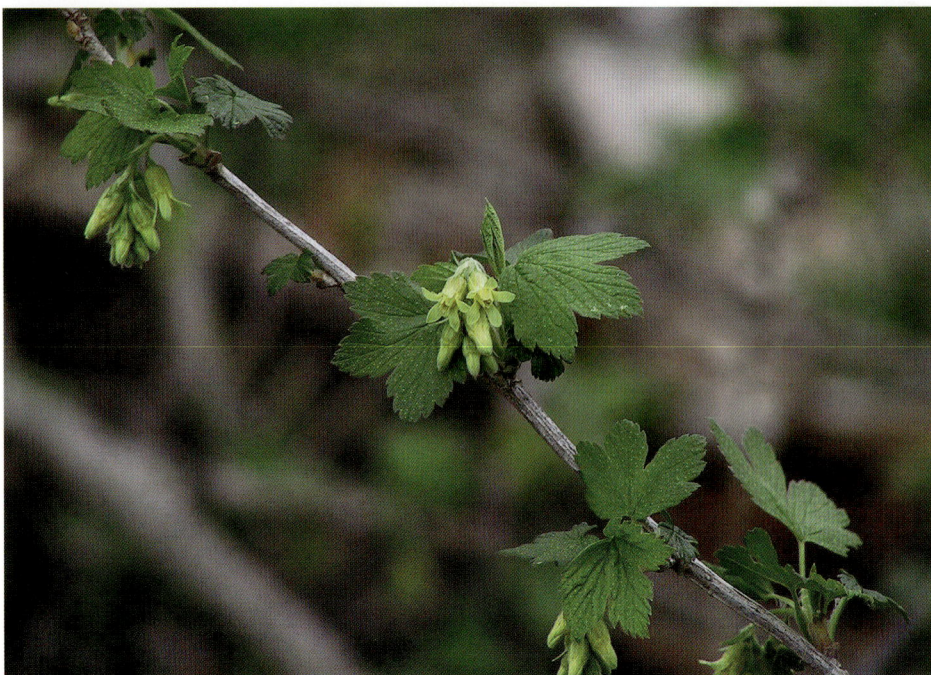

WILD BLACK CURRANT
Ribes americanum
CURRANT FAMILY *(Grossulariaceae)*

Other Common Names: American wild currant; wild currant.

Flowering Period: Apr – May.

Occurrence: Moist ravines and along streams. **FF:** Rare along South and North Stream Trails.

Description: This native shrub grows up to 6' tall. Its branches are straight and gray. The leaves, up to 3" long and wide, have 3 lobes. They appear in clusters along the branches, along with bundles of yellowish, dangling tube flowers. The fruit is a spherical berry about 3/8" in diameter; it turns from green to red, and finally to black by late August.

Identification: The closely related wild gooseberry *(Ribes missouriense)* has similar leaves, but white skinny tube flowers, prickles and purple fruit.

WILD GOOSEBERRY
Ribes missouriense
CURRANT FAMILY *(Grossulariaceae)*

Flowering Period: Apr – May.

Occurrence: Upland and floodplain woods.
FF: Common along Hawthorn Trail. **NW:**
Common along all woodland trails.

Description: This native shrub, with arching
branches, grows up to 5' tall. Its small
leaves are lobed and rounded; they are
usually clustered at the branch nodes and
accompanied by 1-4 red-brown spines. The
flowers are greenish-white and slim; they
dangle from the twigs. The fruit is a
spherical berry which turns from green to
purple at maturity.

Identification: The related wild black currant
(Ribes americanum) has yellowish flower
clusters and lacks spines. Its fruit is black at
maturity.

BLACK RASPBERRY
Rubus occidentalis
ROSE FAMILY *(Rosaceae)*

Flowering Period: Apr – Jun.

Occurrence: Open woods and hillsides. **FF:** Uncommon along Stream Trail. **NW:** Uncommon near the Krimlofski Tract entrance.

Description: This native shrub has purple stems with a white, powdery "bloom" and hooked prickles. The leaves have 3 or 5 lance-shaped leaflets, green above and silvery below. The flowers, with 5 white petals, form clusters of 3-9. The ripe fruit is a black hemispheric berry.

Comments: Edwin James, botanist of the Long Expedition, collected the original (type) specimen of this plant near Neale Woods in 1820.

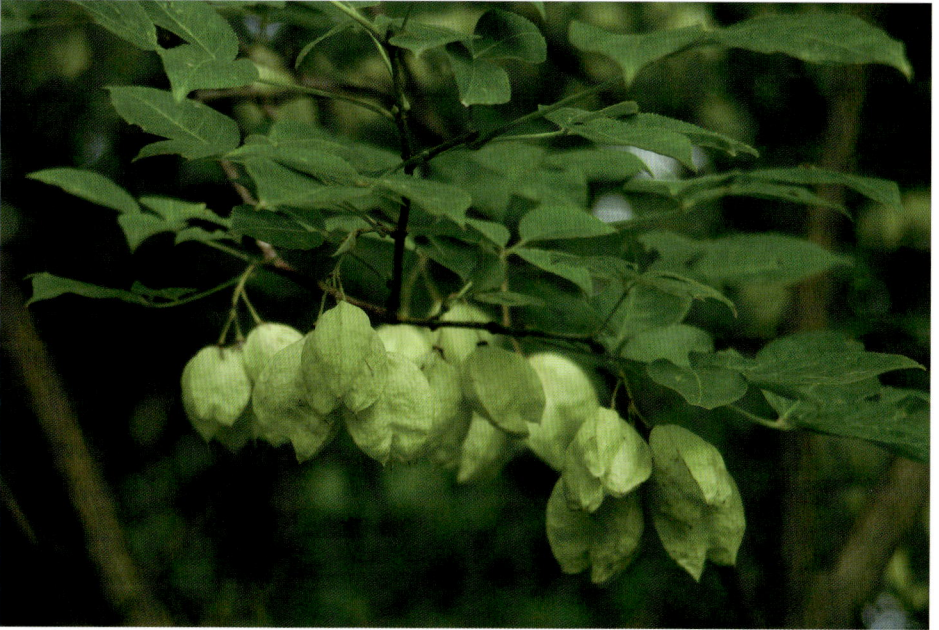

BLADDERNUT
Staphylea trifolia
BLADDERNUT FAMILY *(Staphyleaceae)*

Flowering Period: Apr – May.

Occurrence: Woodlands, especially near water. **NW:** Locally common on Pond Trail. **FF:** Bladdernut was common in FF 15 years ago, but none has been seen along trails recently.

Description: This native shrub grows up to 10' tall, often forming thickets from root suckers. The leaves, which emerge with the flowers, each have three pointed, oval leaflets, up to 4" long. The white, bell-shaped flowers form drooping clusters. The fruit is a 3-chambered, inflated bladder, 1-2" long, first yellowish green, then dark brown at maturity in September. The brown bladders may be seen on this plant throughout the winter and into early spring.

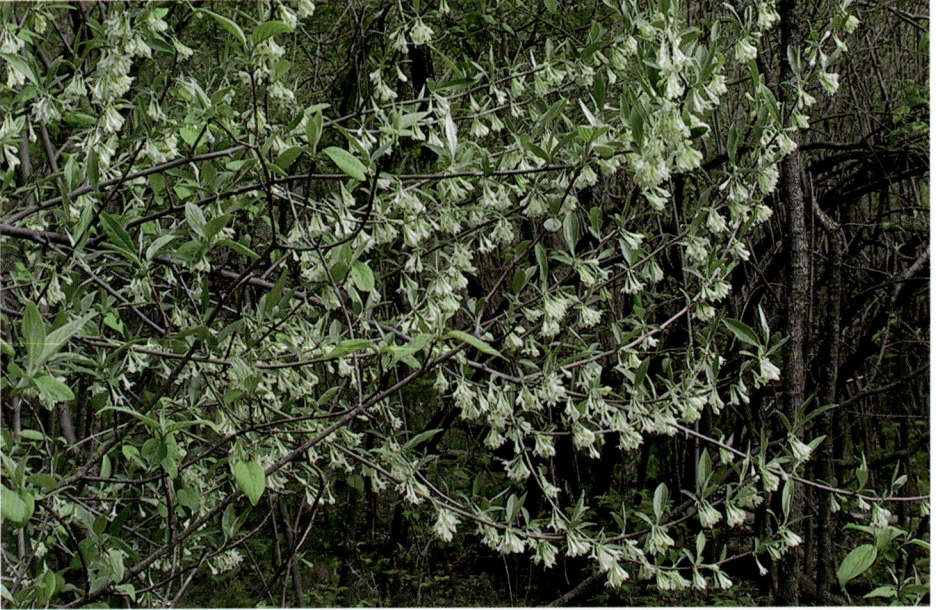

AUTUMN OLIVE
Elaeagnus umbellata
OLEASTER FAMILY *(Elaeagnaceae)*

Flowering Period: Apr – May.

Occurrence: In all types of soil; its roots can fix their own nitrogen. **FF:** Rare along Stream Trail. **NW:** Abundant along Neale Trail below the overlook.

Description: This highly invasive shrub, first introduced from East Asia, can grow up to 15' tall. Its oval, pointed leaves are alternate on straight branches and twigs. Leaf undersides are silvery white. The abundant, light yellow tube flowers dangle in clusters from the stems. Mature shrubs carry large quantities of red berries by July. This shrub is out of control at NW.

Identification: The closely related Russian olive *(Elaeagnus angustifolia)* has yellow flowers and tan fruit.

WAHOO
Euonymus atropurpurea
STAFF TREE FAMILY *(Celastraceae)*

Other Common Name: Spindle tree.

Flowering Period: May – Jun.

Occurrence: Stream banks and wooded areas. **FF:** Uncommon along Stream Trail.

Description: This native shrub or small tree may grow up to 12' tall. The lance-shaped leaves, opposite on straight branches, are up to 4" long. The striking maroon flowers with 4 petals appear in numerous loose clusters. The pink fruit capsule splits by October to reveal up to 4 shiny red seeds. This fruit may often be seen on the shrub well into winter. Its fall foliage is a deep reddish purple.

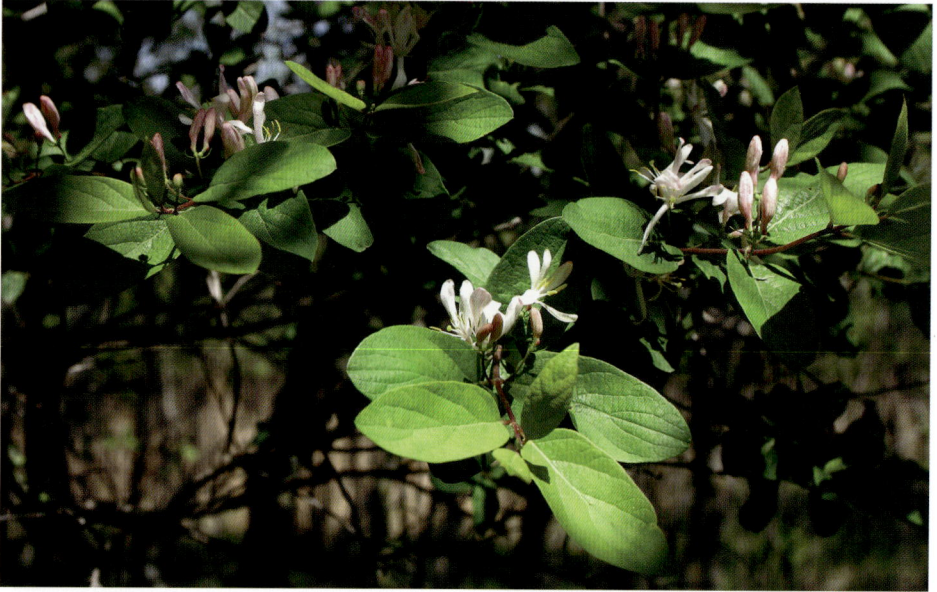

TATARIAN HONEYSUCKLE
Lonicera tatarica
HONEYSUCKLE FAMILY *(Caprifoliaceae)*

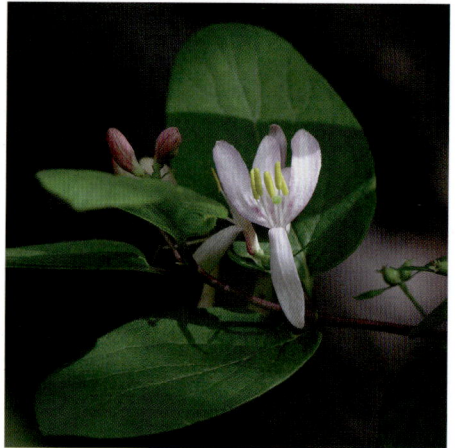

Flowering Period: Apr. – May.

Occurrence: Open woods, stream banks and where planted. **NW:** Rare; scattered around parking lot, and on edges of Nebraska and Knull Prairies.

Description: This naturalized shrub or small tree has leaves which are opposite on branches, elliptic and with rounded tips. The white-to-pink flowers stand vertically in pairs from the leaf nodes. The fruit, an orange berry, ripens in June.

Identification: Wild honeysuckle *(Lonicera dioica)*, our only native honeysuckle, is a vine. Maack honeysuckle *(Lonicera maackii)*, another introduced species, has pointed leaves and red berries.

MAACK HONEYSUCKLE
Lonicera maackii
HONEYSUCKLE FAMILY *(Caprifoliaceae)*

Other Common Name: Amur honeysuckle.

Flowering Period: May – Jun.

Occurrence: On floodplains near water. **FF:** Uncommon along Stream Trail. **NW:** Uncommon on MRE Trail.

Description: This non-native shrub or small tree grows up to 15' tall. The sharply pointed, lance-shaped leaves are opposite on the twigs, with 2 pairs of white, upright flowers at each leaf node. Its shiny red berries appear in September. Leaves and fruit remain on these shrubs late into fall, making them easy to spot after the other shrubs have already lost their leaves.

Identification: Tatarian honeysuckle *(Lonicera tatarica)* has rounded leaves and orange berries.

Comments: This invasive plant is being controlled by cutting.

FALSE INDIGO
Amorpha fruticosa
BEAN FAMILY *(Fabaceae)*

Flowering Period: May – Jun.

Occurrence: Various moist habitats, including prairies. **NW:** Uncommon on Knull Prairie.

Description: This native shrub grows to 10' tall, usually with several stems from its base. It spreads by seed or lateral root sprouts. The compound leaves have up to 20 oval leaflets arranged alternate, but sometimes opposite, on the leaf stalk. The deep purple flowers with bright orange stamens are crowded on erect clusters (racemes). They are sweetly fragrant and provide a striking color display when in bloom. The fruit is a sticky, greenish brown legume (right).

COMMON BUCKTHORN
Rhamnus cathartica
BUCKTHORN FAMILY *(Rhamnaceaea)*

Flowering Period: May.

Occurrence: Along roadsides and in woodlands. **FF:** Uncommon along South Stream Trail.

Description: This dense, non-native shrub or small tree grows up to 15' tall. Its <u>oval, toothed leaves, about 3" long and 2" wide,</u> are mostly opposite on the twigs. The long thorns at the ends of twigs may best be seen after the leaves have fallen. Female and male flowers are on the same plant; the image at right shows the greenish yellow male flowers. The <u>spherical</u> <u>fruit, about ½" in diameter, is first green then black</u>.

Comments: This invasive shrub is being controlled by cutting. It is host to oat crown rust, the orange spots in the images above and right.

FRAGRANT SUMAC
Rhus aromatica
CASHEW FAMILY *(Anacardiaceae)*

Other Common Name: Polecat bush.

Flowering Period: Apr – May.

Occurrence: NW: Rare along southwest edge of Knull Prairie; probably planted there.

Description: This dense native shrub has fine hairs on its young branches, leaves and fruit. Its leaves have three lobed leaflets (trifoliate). The yellowish flowers form dense terminal clusters or spikes. The fruit clusters turn red in June; they are covered with dense hairs.

Comments: The leaves are fragrant when bruised, reflecting its scientific, as well as one of its common names.

SMOOTH SUMAC
Rhus glabra
CASHEW FAMILY *(Anacardiaceae)*

Flowering Period: May – Jun.

Occurrence: Woodland borders, prairies and waste places. **NW:** Common on Jonas and Nebraska Prairies.

Description: This shrub forms dense thickets up to 12' tall. It spreads via root suckers to form large colonies. Its leaves have 11-25 lance-shaped leaflets, opposite on purplish stalks. Male and female flowers are found on separate plants in dense clusters; both are pale greenish (male at right). The fruit forms large, upright spikes of red-brown drupes. The leaves turn a bright reddish purple in the fall.

Comments: Smooth sumac likes to invade prairies. Natural prairie fires would have kept such invasions in check. On NW prairies this plant is controlled by cutting and controlled burning.

MUTLIFLORA ROSE
Rosa multiflora
ROSE FAMILY *(Rosaceae)*

Flowering Period: May – Jun.

Occurrence: Along roadsides, creeks and open woodlands. **FF:** Uncommon along Camp Gifford Road. **NW:** Rare on MRE Trail.

Description: This fast-growing, non-native shrub stands alone or sprawls over other vegetation up to 15' high. The twigs are armed with sharply recurved prickles. Its compound leaves have 7 or 9 oblong leaflets. The flowers, arranged in dense clusters, have 5 white petals and yellow stamens. The fruit is a red rose hip.

Comments: This aggressive shrub is being controlled at both of our nature centers by cutting and up-rooting.

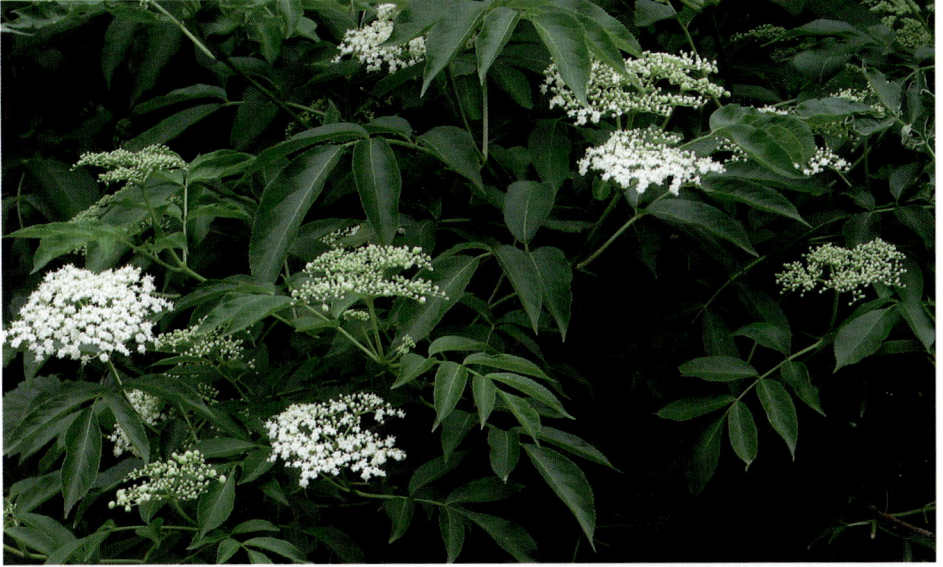

COMMON ELDERBERRY
Sambucus canadensis
HONEYSUCKLE FAMILY *(Caprifoliaceae)*

Other Common Names: American elderberry, elderberry.

Flowering Period: May – Jun.

Occurrence: Moist woods, stream banks.
FF: Uncommon along Camp Gifford Road.
NW: Uncommon along MRE Trail.

Description: This native shrub, up to 15' tall, has a dark brown bark with lighter, raised pimples. The compound leaves have 7 or 9 pointed leaflets with small, sharp teeth. The large, flat-topped flower clusters (cymes) are fragrant. Each flower has 5 creamy-white lobes. The fruit clusters, with purple-to-black berries, mature in July and August.

Comments: Elderberries are used in making preserves, jellies and wine.

HIGHBUSH CRANBERRY
Viburnum opulus
HONEYSUCKLE FAMILY *(Caprifoliaceae)*

Flowering Period: May.

Occurrence: Moist wooded hillsides.
FF: Rare; several shrubs (possibly planted) near entrance to Mill Hollow.

Description: This bushy shrub grows up to 15' tall. The 3-5" leaf blades have 3 lobes. There are two types of flowers, arranged in flat clusters (cymes) about 5" across. The large white 5-lobed flowers on the margins are sterile. The smaller cream-colored flowers in the middle are fertile; they later produce bright red fruit which ripens in August. Winter twigs have opposite, greenish buds.

Identification: The common elderberry *(Sambucus canadensis)* also has flat white flower clusters but only one type of flower; and it has compound leaves.

CORALBERRY
Synphoricarpos orbiculatus
HONEYSUCKLE FAMILY *(Caprifoliaceae)*

Other Common Name: Buckbrush.

Flowering Period: Jul – Aug.

Occurrence: Upland and floodplain woods.
FF: Common along Hidden Lake Trail. **NW:**
Common along History Trail.

Description: This native shrub grows up to
5' tall. It often spreads from its roots to form
large colonies. Its elliptic leaves are opposite
on straight twigs. The greenish flower
clusters hang from the leaf nodes. Its
spherical fruit ripens by late September to
form clusters of coral-colored berries. Where
not grazed by deer in the fall, the fruit may
be seen well into winter.

WINGED BURNING BUSH
Euonymus alatus
STAFF TREE FAMILY *(Celastraceae)*

Description: This naturalized shrub was introduced from Northeast Asia as an ornamental for its bright purplish red fall foliage. It grows as a shrub or small tree up to 15' tall and is considered invasive. One shrub was found in Raccoon Hollow at NW. Its lance-shaped leaves are opposite on the twigs. The small yellowish green flowers bloom in May. The stems have unusual, corky wings on both sides. Its bright red fruit is an elongated drupe, which can be seen on the shrub in September and October.

SAND SAGEBRUSH
Artemisia filifolia
SUNFLOWER FAMILY *(Asteraceae)*

Description: This solitary shrub, a pleasantly aromatic perennial, was probably planted along with the other prairie plants in Jonas Prairie at NW. It normally grows only in Western Nebraska and throughout the Southwest, at elevations up to 6000'. Also known as sagebrush, silver sage and sand sage, it blooms with small yellow flowers in August and September. Its very thin, silvery leaves give this shrub a bluish-green appearance. Native Americans brewed a tea with this plant for a variety of medicinal purposes to include the treatment of stomach and other digestive ailments.

GREENBRIAR
Smilax hispida
CATBRIER FAMILY *(Smilacaceae)*

Flowering Period: Apr - May.

Occurrence: Thickets and woodlands, especially near water. **FF:** Common along Stream Trail. **NW:** Common along MRE Trail.

Description: This native vine climbs over shrubs and young trees; it attaches with tendrils. The leaves are alternate on the vine, oval, pointed and variable in size and up to 9" long on vigorous vines. Male and female flowers are found on separate plants. Both are yellowish green and form spherical clusters. The larger vines are armed with dense, black, dagger-like prickles of various sizes. The berry-like fruit is found in tight clusters, first shiny green, later black. The green leaves and fruit may be seen on the vine late into fall.

Comments: Avoid getting in contact with this vine; its prickles can cause serious damage to skin.

RIVER-BANK GRAPE
Vitis riparia
GRAPE FAMILY *(Vitaceae)*

Flowering Period: May – Jun.

Occurrence: Floodplain woodlands, near water. **FF:** Common on Stream Trail. **NW:** Common around the CJIC.

Description: This native vine climbs over shrubs and trees up to 80' high. Young branches are usually reddish. The tendrils are opposite on the stem to coarse-toothed leaves, which are about as wide as they are long. Elongated flower clusters (thyrse) produce crowded bunches of purple grapes with a powdery, white coating (bloom) by late July. The bark of the mature vine has loose, ropy shreds.

Identification: The similar porcelain berry *(Ampelopsis cordata)* blooms and climbs over the same hosts a little later in spring. Its mature vine lacks the ropy shreds; its fruit is a cluster of orange-pink or turquoise-blue berries.

PORCELAIN BERRY
Ampelopsis cordata
GRAPE FAMILY *(Vitaceae)*

Other Common Name: Raccoon grape, heart-leaf peppervine.

Flowering Period: Jun – Jul.

Occurrence: Mainly floodplain woods, especially near water. **FF:** Locally common along Camp Gifford Road. **NW:** Uncommon on River Loop Trail.

Description: This high-climbing native vine has variable but often heart-shaped leaf blades up to 6" long. It competes with other vines to drape over shrubs and trees. The mature vine, up to 4" in diameter, resembles a small tree trunk. The small greenish flowers are found in clusters (cymes). The fruit is a round berry, which turns from orange-pink to turquoise-blue.

Identification: The similar river-bank grape *(Vitis riparia)* flowers and climbs over the same host plants somewhat earlier in spring. It has purple grapes; its mature vines have a shaggy bark.

POISON IVY
Toxicodendron radicans
CASHEW FAMILY *(Anacardiaceae)*

Flowering Period: May – Jun.

Occurrence: In upland and floodplain woods. **FF:** Common on Hidden Lake Trail. **NW:** Abundant on River Trail.

Description: This native vine climbs tall trees or scrambles along the ground. Its vines may be smooth or have aerial roots (lower right). Mature vines may reach 5" in diameter. The leaves have three pointed leaflets attached to stalks of variable length. The flowers, green with orange stamens, are arranged in clusters (thyrse). The small fruit is spherical and gray (middle right). By October the leaves turn orange or wine-red.

Comments: All parts of this plant may produce a severe contact allergy. Leaflets of three – let it be!!!

WILD HONEYSUCKLE
Lonicera dioica
HONEYSUCKLE FAMILY *(Caprifoliaceae)*

Flowering Period: May – Jun.

Occurrence: Wooded hillsides and along streams. **FF:** Rare; look along History Trail overlooking Spring Hollow.

Description: This native vine trails over other vegetation and down inclines. Its leaves are ovate and opposite on the stem. The uppermost pair of leaves surrounds the stem (perfoliate). These leaves frame a cluster of pale yellow or rose-colored flowers. The fruit is a red berry, which ripens by mid-summer.

Identification: Tatarian honeysuckle *(Lonicera tatarica)* and Maack honeysuckle *(Lonicera maackii)*, the other two honeysuckles found in our two Nature Centers, are shrubs or small trees. Neither have perfoliate leaves.

MOONSEED
Menispermum canadense
MOONSEED FAMILY *(Menispermaceae)*

Flowering Period: May – Jun.

Occurrence: Damp woods. **FF:** Climbing plants are locally common on the fence around the deer exclosure on North Stream Trail. Ground plants are common along most woodland trails. **NW:** Uncommon along Columbine Trail.

Description: This native vine grows along the ground or climbs other plants and fences up to 12' high. Its vine lacks tendrils; it climbs by wrapping itself tightly around other plants and fences. The leaves have up to 6 pointed lobes. A typical ground plant, which usually does not produce flowers, is shown at upper right. Male and female flowers form drooping clusters on separate climbing plants (male at top and left, female right). The spherical fruit, about ½" in diameter, first green (shown at right), turns dark blue late in the fall. It is rarely seen in our area.

AMERICAN BITTERSWEET
Celastrus scandens
STAFF TREE FAMILY *(Celastraceae)*

Flowering Period: May – Jun.

Occurrence: Woodland hillsides and near water. **FF:** Rare along Stream Trail. **NW:** Uncommon at entrance to Knull Prairie and elsewhere.

Description: This native vine sprawls over other vegetation but may also have a primary woody stem of up to 1" in diameter. It spreads by root suckers as well as by seed. The leaves are alternate on the vine, elliptic and pointed. The small greenish flowers are arranged in clusters. The berry-like fruit turns from yellow to orange and splits by October to expose scarlet seeds which may persist on the vine into winter.

VIRGINIA CREEPER
Parthenocissus quinquefolia
GRAPE FAMILY *(Vitaceae)*

Flowering Period: May – Jun.

Occurrence: Open woodland. **FF:** Common along Cottonwood Trail. **NW:** Common along woodland trails.

Description: This native vine climbs tall trees or scrambles along the ground. The dull green leaves are palmately compound, usually with 5 coarsely toothed leaflets. The young vines and leaves are hairy. The branched tendrils end in adhesive disks. Only climbing plants bear flowers and fruit. The yellow-green flower clusters are inconspicuous and found higher up on the vine. The spherical fruit is dark blue.

Identification: The similar woodbine *(Parthenocissus vitacea)* lacks the adhesive disks on the tendrils, and has smooth, shiny young leaves.

WOODBINE
Parthenocissus vitacea
GRAPE FAMILY *(Vitaceae)*

Flowering Period: May – Jul.

Occurrence: Open woods, especially near water. **FF:** Rare along the southern end of South Stream Trail.

Description: This native vine climbs over shrubs and other low vegetation with <u>twining tendrils</u>. The mature leaves are palmately compound with 5 <u>shiny leaflets</u>. Its yellowish flowers form loose clusters. The fruit, on bright red stalks, resembles grapes.

Identification: The closely related Virginia creeper *(Parthenocissus quinquefolia)* has tendrils with adhesive disks and usually dull upper leaf surfaces. Riverbank grape *(Vitis riparia)* has similar fruit but very different leaves.

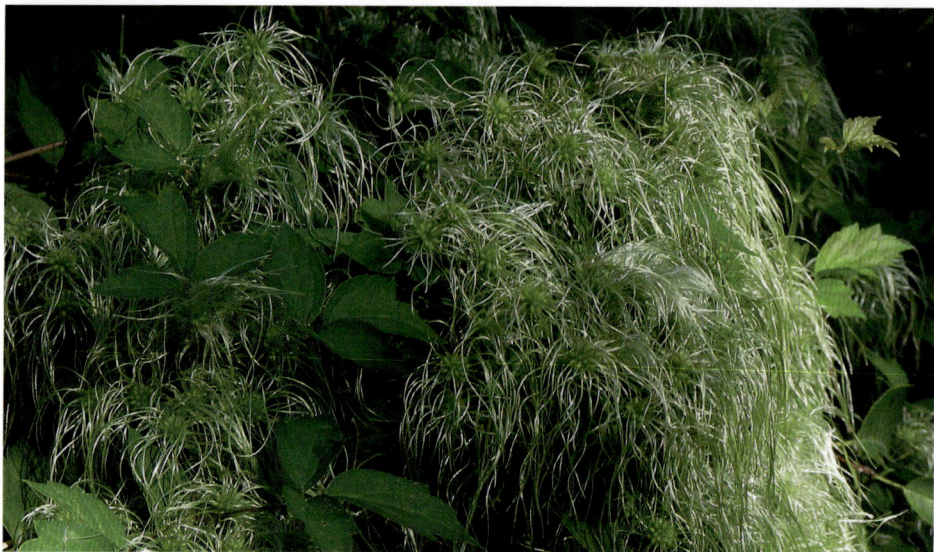

VIRGIN'S BOWER
Clematis virginiana
BUTTERCUP FAMILY *(Ranunculaceae)*

Flowering Period: Jul – Aug.

Occurrence: Roadsides, wet ditches. **FF:** Rare; look along Camp Gifford Road, just east of where it crosses the railroad tracks.

Description: This native vine has male and female flowers on separate plants (dioecious). The female plant is shown above, scrambling over other vegetation after going to seed by late August. The male flower, similar to the female one, is shown at right. It is about 1" across and has 4 slender, creamy white sepals. The compound leaves have three lobed and coarsely toothed leaflets. The dried seeds, attached to long plumes, form wispy clusters (lower right).They persist on the vine throughout the winter.

Comments: This is the native relative to the many varieties of clematis planted in our local gardens.

Sedges and Rushes

This second part features accounts of 32 sedges in 3 genera; 27 from the genus Carex, 2 bulrushes (genus Scirpus) and 3 umbrella or nutsedges (genus Cyperus). Descriptions of 2 rush species follow the sedge accounts.

Sedges are grasslike native perennials and are among our earliest plants to flower, many at the same time as our early woodland wildflowers. Unlike the grasses, they have **triangular, solid stems** which have **no joints** (nodes) and leaves with **tubular, undivided sheaths** (the portion of the leaf that wraps around the stem). **Leaves are 3 ranked** arising from all 3 sides of the stem.

Most of our sedges are in the **genus Carex** whose female flowers are enclosed within a sac-like structure unique to this genus, the perigynium. Flowers are unisexual, but male and female flowers are usually present on the same plant. Flower clusters are arranged in **spikes** which may be all male, all female or mixed (see Illustrated Glossary). If mixed, the spikes may have either male or female flowers at the tip. Male flowers have 3 stamens and a scale extending upward from the base of each flower. Female flowers have a single ovary with a style and 2 or 3 stigmas surrounded by the unique sac-like **perigynium**, which has the following external features often used for identification. The **body** of the perigynium has, in most cases, an extension of variable length known as the **beak**, which is often **toothed** at its tip. There is an opening at the tip of the beak from which 2 or 3 **stigmas** may protrude. A **female scale** arises at the base of each perigynium extending upward a variable distance alongside it. The fruit falls at maturity still enclosed in the perigynium.

The bulrushes (genus Scirpus) and umbrella or nutsedges (genus Cyperus) have inconspicuous bisexual flowers. Our bulrushes are large, stout-stemmed wetland plants whose flower clusters are cylindrical, cone-shaped or arranged in globelike bunches. Our nutsedges are shorter (less than 2 1/2'), often weedy plants of wet areas with elongated brush-like flower clusters made up of many individual spikelets coming off at right angles to its stem.

Rushes are grasslike plants with **solid, round stems without joints** and, leaf sheaths at least partially open. They have tiny, lily-like **flowers with 3 inconspicuous petals and 3 similar sepals** arranged in a circle around the reproductive parts. Their **fruit is a capsule filled with many tiny seeds.**

Tips on using this part of the book

- Take a few minutes to learn the unique characteristics of sedges and rushes and a few basics about the genus Carex that are highlighted above and illustrated in the glossary at the back of the book.
- Plants in the genus Carex are presented first, roughly in the order of their flowering dates.
- Bulrush, umbrella sedge and rush accounts follow the Carex species.
- The habitat in which the plant occurs at FF/NW and unique identifying characteristics are underlined in the species accounts.
- A table categorizing Carex species by habitat and easily identifiable differences in the flower clusters (spikes), which should narrow the number of choices to 10 or less, is located at the end of this section.
- A hand lens will be necessary to identify the smaller sedge features described in the following pages.

BLUNT-SCALED OAK SEDGE
Carex albicans
SEDGE FAMILY *(Cyperaceae)*

Flowering Period: Late Mar - May.

Occurrence: Moist to dry <u>upland woods</u>, especially upper south-facing slopes. **FF:** Common at Hackberry/Wren Trail junction. **NW:** Common on Fox and upper Settlers Trails.

Description: <u>One of our earliest flowering plants</u>, this sedge forms single or loosely grouped, dense <u>cushion-like bunches</u>. Early flowering stems up to 15" tall are erect and nearly all the same height, rising well above the narrow 1/8" wide leaves. Stems droop at maturity with the tips often touching the ground. The compact flower cluster has a <u>separate male spike</u> with yellow anthers at the tip and 1-4 sessile female spikes with white stigmas below (lower left). Mature plants have finely hairy, toothed perigynia with prominent beaks, and <u>scales with dark reddish-purple or brown margins</u> (lower right).

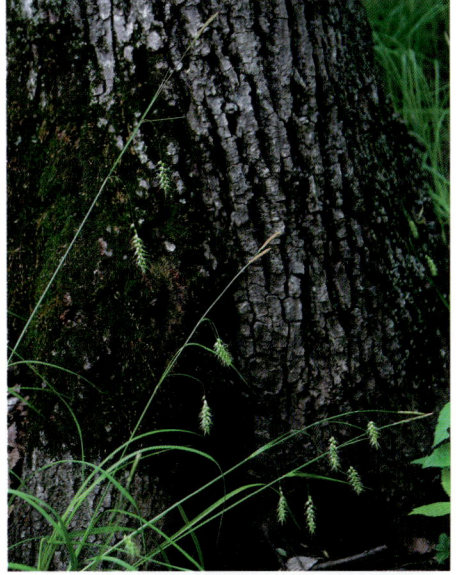

LONG-BEAKED SEDGE
Carex sprengelii
SEDGE FAMILY *(Cyperaceae)*

Flowering Period: Mid Apr - May.

Occurrence: Upland and floodplain woods.
FF: Common on Hackberry Trail. **NW:**
Abundant on Hilltop and upper Bittersweet
Trails.

Description: Rhizomatous native perennial
forming dense, sometimes donut-shaped
clumps (lower right) or extensive colonies
like the one at NW on Bittersweet Trail
shown above. Flowering stems from 12-30"
tall are about the same length or slightly
longer than the 1/8-1/4" wide leaves. At the
top of the stem are 1-3 male or mostly male
spikes; below are 2-4 separate long-
stemmed, female spikes which droop when
mature. The mature light green to straw-
colored perigynia have globular bases and
characteristic long, thin beaks (middle,
right), as the common name suggests.

WOODLAND SEDGE
Carex blanda
SEDGE FAMILY *(Cyperaceae)*

Flowering Period: Mid Apr - Jul.

Occurrence: <u>Many habitats</u> including wooded bottomlands, upland woods, streambanks, ditches, lawns, disturbed areas. **FF and NW:** Abundant along most woodland trails.

Description: Perennial growing in well-defined bunches. Sprawling dull green to grayish green, fairly broad, grooved leaves up to 3/8" wide give some plants a "crabgrass-like" appearance, especially early in the season (upper left). Later they often assume a more conventional look (upper right). <u>Sharply triangular flowering stems</u> from 4-24" tall, many with narrow wings at the angles, (lower left) bear a <u>separate male spike at the tip</u> and up to 4 erect female spikes below. Observation with a hand lens will show that the <u>tips or beaks of the mature perigynia are curved</u> (lower right).

Identification: Sharply triangular flowering stems and perigynia with curved beaks separate *C. blanda* from gray wood sedge *(C. grisea)* and few-fruited sedge *(C. oligocarpa)*. Hitchcock's Sedge *(C. hitchcockiana)* has perigynia with beaks bent to one side, but they are more widely spaced, and leaf sheaths are hairy and usually rough to the touch.

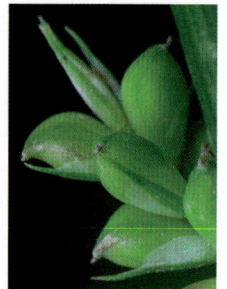

Comments: This is our most common and widely distributed sedge, thriving in both floodplain and upland habitats.

HITCHCOCK'S SEDGE
Carex hitchcockiana
SEDGE FAMILY *(Cyperaceae)*

Flowering Period: Late Apr - May.

Occurrence: <u>Moist woodlands</u>, especially at the base of bluffs or lower slopes of wooded ravines. **FF:** Common in Handsome Hollow. **NW:** Common on lower Paw Paw Trail and woodland section of Jonas Trail.

Description: Bunched perennial with flowering stems 6-24" tall bearing a separate <u>male spike at the tip</u> and 3-5 erect female spikes below. Each spike has only a <u>few</u> (usually 2-7) <u>loosely spaced perigynia</u> whose <u>beaks are bent</u> to one side. The leaf <u>sheaths</u> usually <u>have dense, short hairs that are rough to the touch</u>.

Identification: Other sedges of our upland woods with separate male and female spikes do not have hairs on the leaf sheaths. See few-fruited sedge *(C. oligocarpa)* for discussion.

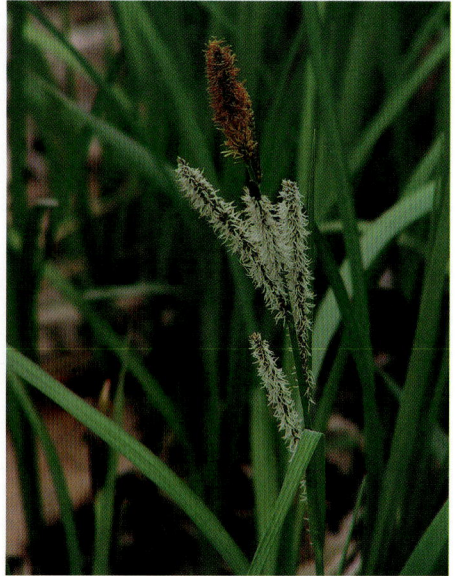

EMORY'S SEDGE
Carex emoryi
SEDGE FAMILY *(Cyperaceae)*

Flowering Period: Mid Apr - May.

Occurrence: <u>Streambanks, wet ditches, marshes</u>. **FF:** Common in both Camp Gifford Road ditches immediately west of parking lot. **NW:** Rare on Missouri River bank near bench on MRE Trail.

Description: Sedge with long rhizomes often forming dense clumps or colonies in wet places. Erect flowering stems 16-48" long are usually taller than the leaves. Each stem bears <u>4-8 spikes</u>, the light brown <u>terminal one always male</u> (upper right). Below, there may be one or more lateral male spikes or mixed spikes with male flowers above the female ones. Lowest spikes are entirely female, the female flowers bearing two showy white stigmas. The <u>flat, green, rounded perigynia have short beaks and no teeth</u>. Scales have reddish brown edges and green midribs (lower right).

Identification: Smoothcone *(C. laeviconica)* and shoreline sedge *(C. hyalinolepis)* have 3 stigmas and long-beaked, toothed perigynia that are circular, not flat, in cross-section.

SAWBEAK SEDGE
Carex stipata
SEDGE FAMILY *(Cyperaceae)*

Flowering Period: Late Apr - Jun.

Occurrence: Marshes, streams, ditches and wet woodlands. **FF:** Rare along Stream Trail and boardwalk in Handsome Hollow.

Description: Bunched perennial with flowering stems 1-4' tall. Overlapping spikes are all alike, consisting of both male and female flowers, the male portions above. The sharply triangular, narrowly winged stems are soft and easily compressed or bent. Broad leaf blades are up to 1/2" wide. Leaf sheaths are loose, the underside thin, white and usually cross-wrinkled, often breaking up at maturity (middle right). Mature spikes bear green to light brown perigynia with long, thin tapered beaks and numerous linear nerves (lower right).

Identification: Soft Fox Sedge *(C. conjuncta)* also has soft, compressible stems and cross-wrinkled sheaths, but perigynia have shorter beaks with fewer, less well-defined nerves. Other species with spikes all alike and superior male flowers do not have spongy, winged stems.

SHORELINE SEDGE
Carex hyalinolepis
SEDGE FAMILY (Cyperaceae)

Flowering Period: Late Apr - Jun.

Occurrence: River floodplains, seasonal wetlands, marshes, wet prairies, ditches. **FF:** Common along edge of Great Marsh and in wet depressions along Hidden Lake Trail.

Description: Sedge with well-developed rhizomes forming large clumps or colonies in wet places. Flowering stems up to 4' tall bear 2-4 male spikes at the top, the scales purplish early in the season, eventually fading to light brown. Immediately below are up to 4 erect or arching female spikes with red-brown scales that fade late in the season. Oval perigynia have long tapered beaks with short teeth (lower right). Long, broad leaves up to 5/8" wide have light brown sheaths, their bases often obscured by the remains of last year's growth (middle right).

Identification: Emory's sedge *(C. emoryi)* has flat perigynia with short beaks and no teeth. Smoothcone sedge *(C. laeviconica)* has narrower, graceful arching leaves and erect female spikes situated well below the males. Stem bases are firmer and rounder and their sheaths are often dissected into laddered fibers.

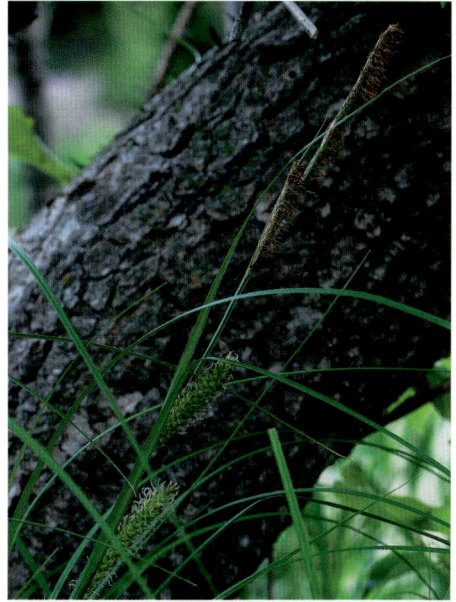

SMOOTHCONE SEDGE
Carex laeviconica
SEDGE FAMILY *(Cyperaceae)*

Flowering Period: Late Apr - Jun.

Occurrence: Wet areas including marshes, streambanks, ditches, lowland woods, wet prairies. **FF:** Common along South Stream Trail. **NW:** Locally common on MRE Trail.

Description: Rhizomatous perennial with flowering stems from 1-4' tall usually extending above the arching, narrow leaves up to 1/3" wide. Above are 2-6 separate brown male spikes with the erect female spikes well below. The rounded, green to yellow perigynia have 3 stigmas and long, tapered beaks with 2 prominent teeth which often diverge slightly (middle right). The reddish tinged female scales fade as the perigynia mature. The firm, rounded lower stem has sheaths that often become dissected into laddered threadlike fibers as they mature (lower right).

Identification: See discussions under Emory's *(C. emoryi)* and shoreline sedge *(C. hyalinolepis)*.

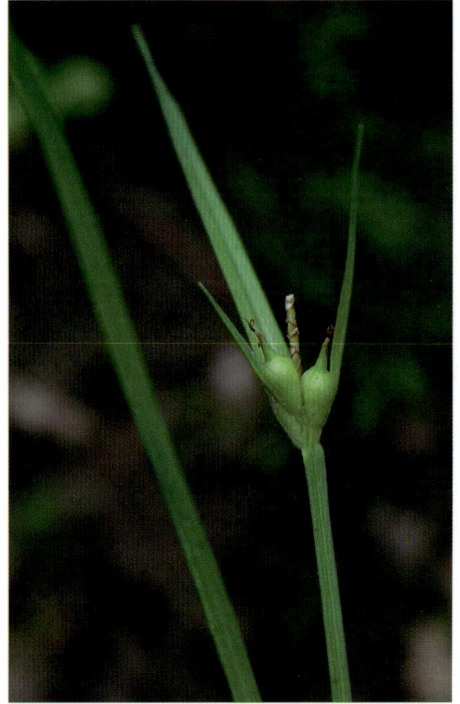

JAMES' SEDGE
Carex jamesii
SEDGE FAMILY *(Cyperaceae)*

Other Common Name: Grass sedge.

Flowering Period: Late Apr - Jun.

Occurrence: <u>Moist bottomland and upland woods.</u> **FF:** Rare on Hawthorn Trail. **NW:** Common on MRE and Settlers Trails.

Description: Bunched perennial with erect or nodding stems 2-12" tall bearing <u>distinctive spikes</u> at their tips. The small central male spike has brown and green scales with white edges and is less than 1/2" tall. The female spike with its 1-4 green perigynia lies at its base (lower right). Perigynia taper abruptly to a long pointed beak. Arising below the perigynia are two narrow 1/2-2" <u>leaf-like bracts which rise above the spikes.</u>

Comments: This species is named for Edwin James, botanist on the 1819-20 Long Expedition. The party wintered just north of NW, but the plant was not collected in that area.

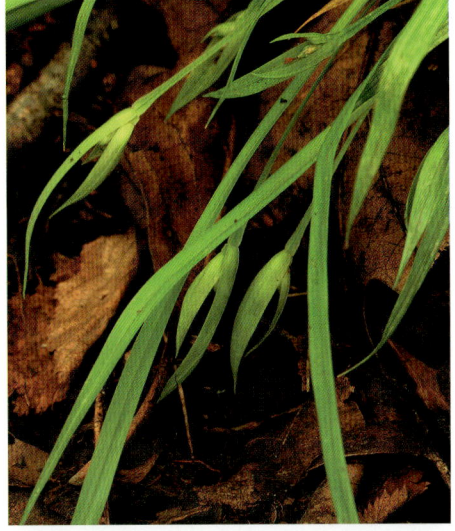

ROCKY MOUNTAIN SEDGE
Carex saximontana
SEDGE FAMILY *(Cyperaceae)*

Flowering Period: Early May - Jun.

Occurrence: Upland woods; moist to dry sites. **FF:** Rare on southern Oak Trail. **NW:** Uncommon with only a few plants at scattered upland sites including Neale Trail just above River Road.

Description: Bunched perennial with flowering stems to 14" tall that are usually shorter than the leaves. Each flowering stem has single male and female spikes at the tip. Additional lower spikes, if present, are basal and on short peduncles. The very short central male spike does not rise above the female spike composed of 2-6 perigynia which surround its base (lower right). Arising just below the perigynia are 2 leaflike bracts which completely envelop the spikes, hiding them from view (upper right).

Identification: James' sedge *(Carex jamesii)* also has two prominent bracts at the base of the male and female spikes, but they are narrower and do not conceal them.

GRAY WOOD SEDGE
Carex grisea
SEDGE FAMILY *(Cyperaceae)*

Flowering Period: Early May - Jun.

Occurrence: Moist upland and lowland woods and ravine slopes. **FF:** Common in Spring and Mormon Hollows. **NW:** Common in Neale/Settlers Trail junction area.

Description: Sedge growing in small bunches. Fairly narrow leaves up to 1/3" wide have smooth and hairless sheaths. Flowering stems from 6-24" tall bear a single separate short-stemmed male spike above and 2-5 erect female spikes below. Female spikes have strongly overlapping perigynia spirally arranged about the long axis. The perigynia are beakless with straight tips.

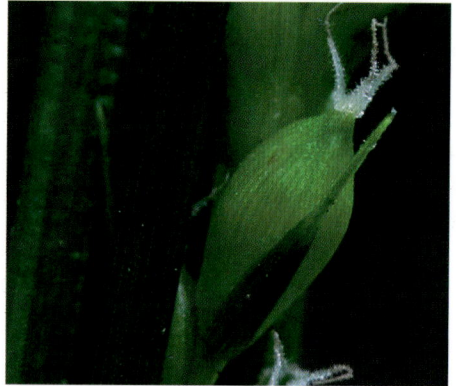

Identification: See discussion under few-fruited sedge *(C. oligocarpa).*

FEW-FRUITED SEDGE
Carex oligocarpa
SEDGE FAMILY *(Cyperaceae)*

Flowering Period: Early May - Jun.

Occurrence: <u>Moist upland woods</u>. **FF:** Uncommon on Hawthorn and Cottonwood Trails. **NW:** Uncommon on Neale Trail above River Road entrance.

Description: Perennial sedge growing in bunches. Narrow leaves up to 1/4" wide have <u>smooth, hairless sheaths.</u> Flowering stems from 6-24" tall bear a short-to-long-stalked <u>male spike at the tip</u>. Below are 2-4 erect female spikes with up to 12 <u>perigynia arranged in 2 opposite ranks.</u> Perigynia are tapered to a <u>short, straight beak</u> at the tip.

Identification: Three other sedges with separate male and female spikes are common in our uplands. Gray wood sedge *(C. grisea)* has female spikes, usually with more perigynia, that are spirally arranged rather than in two opposite ranks. Woodland sedge *(C. blanda)* has perigynia with curved beaks and more triangular, often slightly winged stems. Hitchcock's sedge *(C. hitchcockiana)* has rough hairs on the leaf sheaths and few loosely spaced perigynia with beaks bent to one side.

SLENDER SEDGE
Carex tenera
SEDGE FAMILY *(Cyperaceae)*

Flowering Period: Late Apr - Jun.

Occurrence: Moist floodplain woods. **FF:** Common along GM boardwalk between WLC and the blind. **NW:** Uncommon on MRE Trail.

Description: Bunched perennial with flowering stems 12-30" tall and narrow grass-like leaves. Loosely spaced spikes are borne on slender, arching stems. Spikes are all alike with female flowers above and the often inconspicuous males below (upper). The light green or yellowish perigynia turn a rich golden brown at maturity (lower right).

Identification: At FF/NW, shorter sedge *(C. brevior)* and troublesome sedge *(C. molesta)* have spikes all alike with female parts above, but stems are erect, not arching, and spikes are usually closer together.

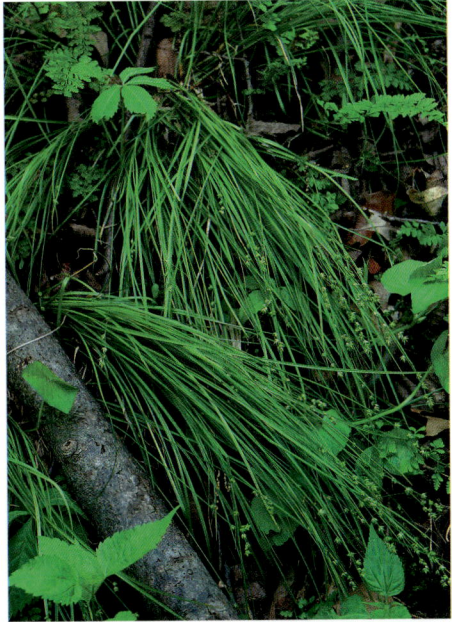

STELLATE SEDGE
Carex rosea
SEDGE FAMILY *(Cyperaceae)*

Flowering Period: Early May - Jun.

Occurrence: <u>Wooded lowlands, moist upland woods and ravines</u>. **FF:** Uncommon in lower Mormon Hollow. **NW:** Common on north facing slopes of Paw Paw and Columbine Trails.

Description: Perennial growing in dense bunches. Early plants are upright (left upper) but they often nod or spread as they mature (right upper). <u>Leaf blades are very narrow,</u> no more than 1/8" wide. Flowering stems from 6-20" tall bear 3-9 spikes, widely separated below but often overlapping above. <u>Spikes are all alike</u> containing central male flowers at the tip and only 5-14 peripheral female flowers. At maturity the <u>perigynia spread out forming star-shaped structures</u>, as the common name suggests (lower right).

Identification: Bur-reed sedge *(C. sparganioides)* has widely spaced spikes, but mature perigynia do not have the characteristic star-shaped or circular arrangement. Its leaves are also much broader.

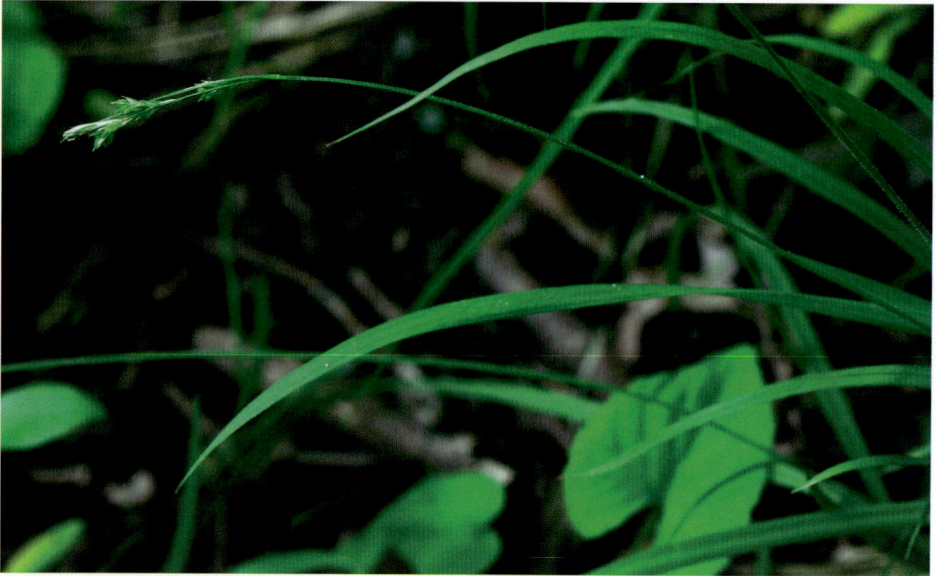

HAIRY WOOD SEDGE
Carex hirtifolia
SEDGE FAMILY *(Cyperaceae)*

Flowering Dates: Apr - Jun.

Occurrence: Moist upland woods, especially lower slopes and ravines; lowland woods. **NW:** Rare on lower Paw Paw Trail.

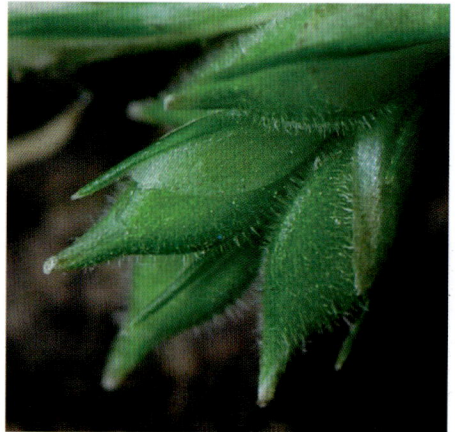

Description: Native sedge growing in small bunches. Erect or arching sharply triangular flowering stems are 8-24" long and hairy along the edges. Leaves up to 15" long are covered with soft hairs. They are basal or arise from the lower 1/3 of the stem. There is a single, separate male spike at the tip of the flowering stem with 2-4 dense to loosely spaced female spikes just below. Each female spike consists of 10-25 green, hairy, triangular perigynia. They are tapered at both ends to a short stalk-like base below and a short beak at the tip.

Comments: The few plants found at a single site in Neale Woods are at the western edge of their range. At present, it is only the second reported occurrence in the state of Nebraska.

MEAD'S SEDGE
Carex meadii
SEDGE FAMILY *(Cyperaceae)*

Flowering Period: Early May - Jun.

Occurrence: <u>Upland prairies</u>. **NW:** Uncommon in Millard Prairie transplant which is located in Koley Prairie at the junction of Jonas and Neale Trails.

Description: Rhizomatous perennial sending up scattered shoots forming open colonies. The 8-20" flowering stem and leaves are grayish-green in color. Spikes are unisexual with a <u>single long-stemmed male spike above</u> and 1-3 female spike(s) below, usually borne on the upper half of the stem. <u>Pale green or yellow brown perigynia</u> have 3

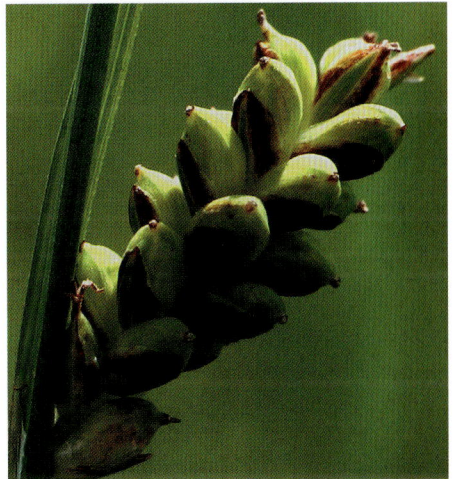

styles and <u>reddish brown scales with a green midrib</u>. The very short beaks at their tips are bent to one side.

HEAVY SEDGE
Carex gravida
SEDGE FAMILY *(Cyperaceae)*

Flowering Period: Early May - Jul.

Occurrence: Openings in upland or lowland woods, open disturbed areas, prairies, roadsides. **FF:** Common on South Stream Trail. **NW:** Common on Krimlofski entrance trail.

Description: Sedge growing in bunches with flowering stems up to 30" tall and leaves up to 12" long and 1/3" wide. The elongate to oval flower cluster is quite compact, bearing 5-10 spikes, overlapping densely at the tip but often slightly separated below. Spikes are all alike, each bearing male flowers at the tip and female ones below (upper middle). Males become inconspicuous after flowering (upper right). Leaf sheaths are thin and transparent at the top (lower left). Sometimes they fit snugly around the stem, but they may be quite loose, baggy and easily torn with prominent cross-wrinkles. The green perigynia turn brown at maturity.

Identification: Glomerate sedge (*Carex aggregata*) is also common at FF/NW. It is not described separately, as many characters overlap, and the authors have had difficulty consistently separating it from heavy sedge. Typically, glomerate sedge prefers shadier sites, the spikes are a little more open and the back of the leaf sheath is white-spotted. Also, the upper margin of the leaf sheath usually is thicker with yellow or brown discoloration as shown in the lower right hand photo (compare with heavy sedge - lower left). Bur-reed sedge (*Carex sparganioides*) is easier to identify. It has broader leaves and an open flower cluster with widely spaced lower spikes.

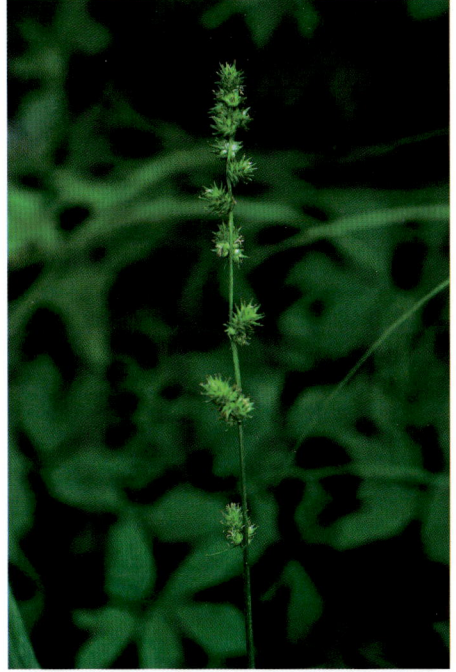

BUR-REED SEDGE
Carex sparganioides
SEDGE FAMILY *(Cyperaceae)*

Flowering Period: May - Jul.

Occurrence: <u>Upland woods</u>. **FF:** Uncommon on lower Linden Trail. **NW:** Uncommon on woodland portion of Nebraska Trail.

Description: Sedge with poorly developed rhizomes growing in bunches. Flowering stems from 16-40" tall bear 6-15 oval spikes. <u>Lower spikes are well separated</u>, but they are closer, often overlapping, at the tip. The rounded to oval <u>spikes are all alike</u> with male flowers, often inconspicuous, above and female flowers below. The green perigynia have just 2 stigmas and short, inconspicuous scales. Arching <u>leaves up to 1/2" wide</u> have loose, baggy and easily torn sheaths (lower) that are whitish or mottled green and white with prominent cross-ridging on the underside.

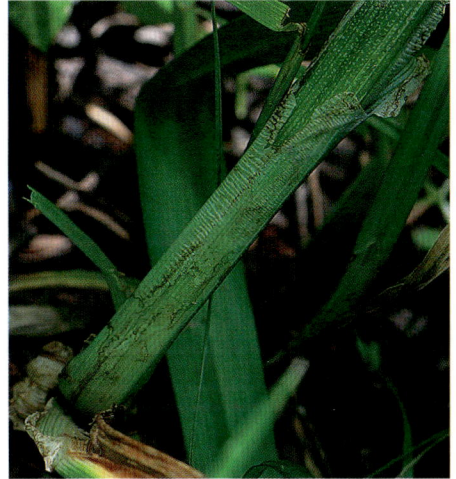

Identification: Heavy sedge *(C. gravida)* and glomerate sedge *(C. aggregata)* have less widely spaced spikes and narrower leaves. They are usually found in more open sites.

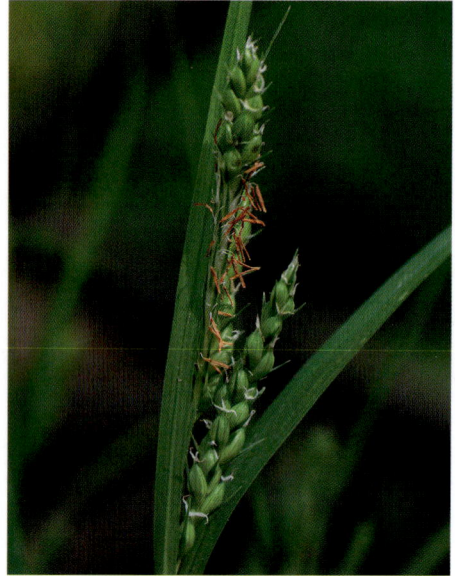

DAVIS' SEDGE
Carex davisii
SEDGE FAMILY *(Cyperaceae)*

Flowering Period: Early May - Jul.

Occurrence: <u>Moist woodlands</u>. **FF:** Common along GM boardwalk. **NW:** Uncommon on MRE Trail.

Description: This attractive bunched sedge has flowering stems from 12-36" tall which may be shorter or longer than the prominent grass-like leaves.   The <u>distinctive upper spike has male flowers below, but the upper portion is female</u> (right upper and middle). Below this bisexual spike are 2-3 <u>female spikes</u> which are erect when young, but usually <u>nod or droop when mature.</u> The green perigynia turn yellow-brown or a striking orange color as they mature (right middle and lower).

Identification: A few of our sedges at FF/NW with predominantly unisexual spikes may have a bisexual upper spike with males at the tip and females below, but none have a mixed upper spike with female flowers at the tip like that of *C. davisii*.

WOODBANK SEDGE
Carex cephalophora
SEDGE FAMILY *(Cyperaceae)*

Flowering Period: Early May - Jun.

Occurrence: Dry to moist <u>upland woods.</u>
FF: Common along Hackberry Trail. **NW:**
Uncommon on wooded portion of Nebraska
Trail.

Description: Sedge growing in bunches
with flowering stems from 8-30" tall, about
the same length as the leaves. From 4-12
densely packed spikes, each with a narrow
hair-like bract at the base, form a <u>compact
oval cluster at the tip of the stem.</u> <u>Spikes
are all alike</u> with inconspicuous male flowers
at the tip and female flowers bearing two
stigmas below. Mature spikes have light
green-to-light brown, flattened perigynia
with minutely toothed beaks.

SOFT FOX SEDGE
Carex conjuncta
SEDGE FAMILY *(Cyperaceae)*

Flowering Period: Early May - Jun.

Occurrence: Moist bottomland woods and prairies, ditches, riverbanks. **FF:** Uncommon along Hidden Lake and Redbud Trails.

Description: Perennial growing in well-defined bunches. Flowering stems are 1-3' tall, sharply triangular often with narrow wings, very soft, compressible and easily bent. Spikes are all alike with male flowers at the top, the males very inconspicuous after flowering. The numerous leaves up to 18" long and 1/3" wide have loose, often wrinkled and easily torn sheaths. Mature perigynia are oval with a relatively short beak.

Identification: Sawbeak sedge *(C. stipata)* has soft, winged stems, but perigynia have longer beaks and prominent linear markings (nerves). Fox sedge *(C. vulpinoidea)* has sharply triangular stems, but they are not winged or soft and spongy.

It also has larger seed heads, smaller perigynia and tight leaf sheaths. Our other sedges with male flowers at the top and spikes that are all alike do not have winged or spongy stems.

FOX SEDGE
Carex vulpinoidea
SEDGE FAMILY *(Cyperaceae)*

Flowering Period: Late May - Jul.

Occurrence: <u>Pond and stream edges, wet meadows</u>, marshes, ditches. **FF:** Uncommon along GM boardwalk. **NW:** Uncommon at edge of pond in Jonas Valley.

Description: Bunched perennial with <u>sharply triangular, firm flowering stems</u> 1-3' tall, often forming dense clumps in wet places. Elongated, <u>dense flower spikes are compound</u>, each side branch consisting of a number of small, separate spikelets. <u>Spikelets are all alike</u>, the inconspicuous male flowers with whitish anthers above and green females below (lower right). Narrow bracts associated with the spikes often give the head a bristly appearance. Narrow, pale green leaves have <u>tight greenish-white sheaths with cross-wrinkled undersides and convex upper margins</u> (lower left). Perigynia are small and turn brown at maturity.

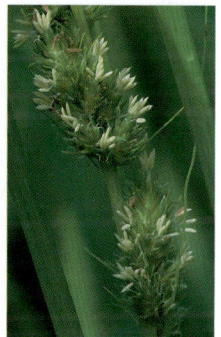

Identification: See discussion under soft fox sedge *(Carex conjuncta)*.

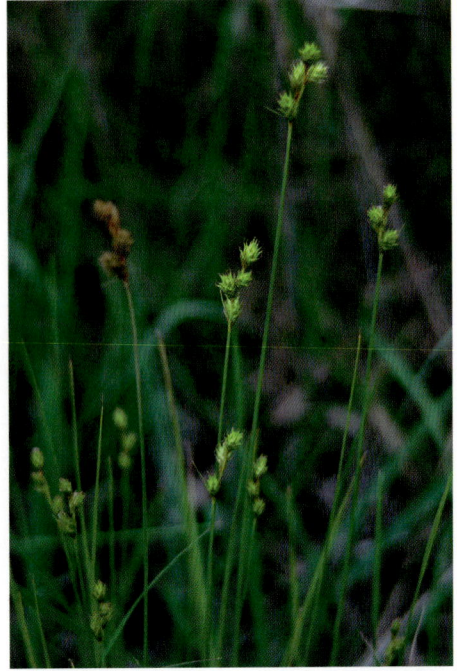

SHORT-BEAKED SEDGE
Carex brevior
SEDGE FAMILY *(Cyperaceae)*

Flowering Period: Mid May - Jun.

Occurrence: <u>Open areas</u> including prairies, pastures, roadsides, open woodlands and low moist sites. **NW:** Common in brome field below Koley Prairie and in Knull Prairie. **FF:** Not documented; most likely overlooked.

Description: Perennial growing in bunches with flowering stems from 6-48" tall, much longer than the late-developing leaves. From 2-6 spikes are clustered at the end of the flowering stem, the lower ones more widely spaced than those at the tip. <u>Individual spikes are all alike</u>, the female flowers situated above the males, which are usually inconspicuous, especially after flowering (lower photos). <u>Spikes</u> are usually tapered at the base and <u>bluntly pointed at the tip</u> (lower right). Perigynia turn golden brown to brown at maturity.

Identification: Typically, troublesome sedge *(C. molesta)* has more compactly arranged spikes that are rounded at the base and apex. The above non-technical features will separate many, but not all, plants in these two very similar species.

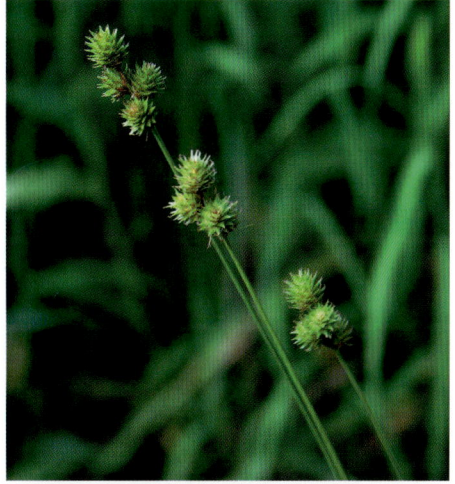

TROUBLESOME SEDGE
Carex molesta
SEDGE FAMILY *(Cyperaceae)*

Flowering Period: Mid May - Jun.

Occurrence: <u>Diverse habitats</u> including moist prairies and woodlands, stream banks, pond margins, ditches and other moist disturbed sites. **FF:** Uncommon along Hidden Lake Trail east of blind. **NW:** Uncommon on Gifford Trail at edge of woodland.

Description: Bunched perennial with flowering stems 10-40" tall rising well above the narrow leaves no more than 1/8" wide. From 2-5 <u>densely overlapping spikes</u> occur at the tip of the stem with the lowermost one often a little more loosely spaced than the others. <u>Individual spikes are all alike</u>, the female flowers situated above the males, which are usually very inconspicuous. <u>Spikes are circular to broadly oval in shape and rounded at both ends</u>. The flattened, winged, green perigynia turn tan or brown at maturity.

Identification: Troublesome sedge closely resembles short-beaked sedge *(Carex brevior)*. In some individuals the non-technical characters used here overlap, making it impossible to separate them in the

field although, typically, short-beaked sedge has more loosely grouped spikes, which are tapered at the ends. Slender sedge *(Carex tenera)* is easier to separate. It has a nodding stem with even more widely spaced spikes than short-beaked sedge. Note that another very similar species in this difficult group, crested sedge *(C. cristatella),* also occurs at FF (rare) and was, perhaps, overlooked at NW.

BOTTLEBRUSH SEDGE
Carex hystericina
SEDGE FAMILY *(Cyperaceae)*

Flowering Period: May - Jul.

Occurrence: <u>Marshes,</u> wet meadows, stream banks, ditches. **FF:** Rare; single clump found next to bridge between Handsome Hollow and the railroad tracks.

Description: Rhizomatous sedge forming clumps in wet places. Leaves are long and narrow, up to 1/4" wide. Flowering stems from 1-3' tall terminate in a <u>single male spike</u> (lower right corner of upper photo). Below are 1-4 <u>nodding, cylindrical female spikes on slender stalks.</u> The egg-shaped perigynia taper abruptly to a long, thin, minutely toothed beak and spread at maturity, giving the spike the <u>bristly appearance</u> responsible for its common name.

Identification: Bearded sedge *(C. comosa)* is more robust with broader leaves and "feathery" perigynia with longer, broadly diverging teeth.

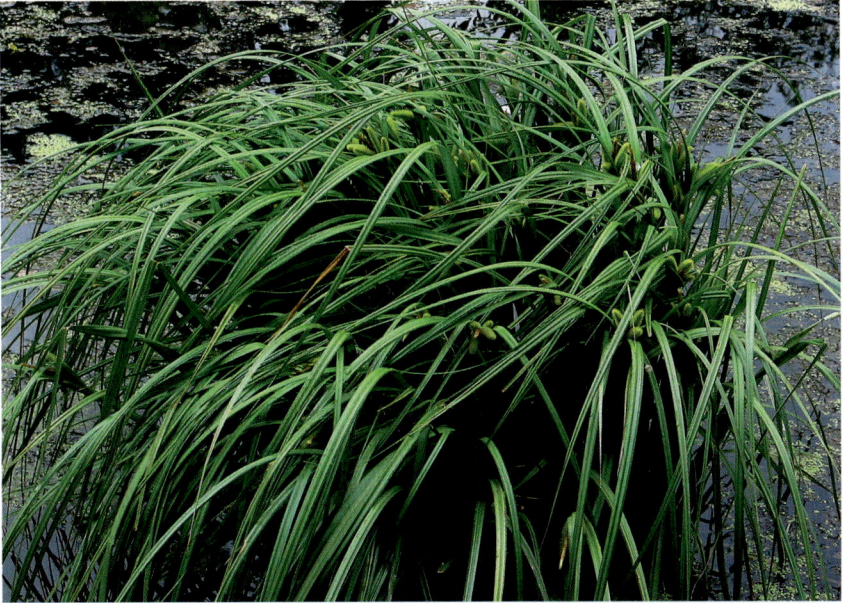

BEARDED SEDGE
Carex comosa
SEDGE FAMILY *(Cyperaceae)*

Flowering Period: Late May - Jul.

Occurrence: <u>Pond and stream margins, often growing in shallow water</u>, wet meadows, bottomland forests. **FF:** Rare; found only along North Stream Trail and in the pond just beyond. The largest plant forms a hummock some 3' in diameter in the middle of the stream.

Description: The last of our sedges to flower, this <u>large sedge</u> often grows in the water forming clumps with flowering stems to 4' tall and broad leaves up to 3/4" wide. At the tip of the stem is a <u>single male spike</u>. Below are 2-7 <u>female spikes</u> borne on slender stalks which <u>droop or nod when mature</u>. The oval-shaped perigynia taper to a <u>prominent beak with two widely spreading teeth</u> resulting in the "fuzzy" look responsible for its common name (lower right).

GRAY'S SEDGE
Carex grayi
SEDGE FAMILY *(Cyperaceae)*

Other Common Names: Globe, bur, mace and morningstar sedge.

Flowering Dates: May - Oct.

Occurrence: <u>Lowland woods</u>, swamps, stream margins. **FF:** Rare and planted in garden area east of WLC.

Description: This sedge, which forms dense clumps, has flowering stems from 1-3' tall with dark green leaves 4-14" long and 1/2" wide. At the tip of the flowering stem is a <u>single narrow male spike</u> on a stalk of variable length. Immediately below are <u>1-2 unique and showy, rounded female spikes</u> up to 1 3/4" in diameter. Spikes consist of 6-35 pale to dark green perigynia whose bases are densely wedged together with the tapered beak-like tips radiating out in all directions. Perigynia are relatively large measuring 1/2 to 7/8" long.

Comments: This plant is native to states immediately to our south and east, but not to Fontenelle Forest. It most likely arrived in seed used to plant the garden area immediately to the east and south of the Wetlands Learning Center. The handsome foliage and unique, showy spikes which persist into the fall and winter have prompted its use in shaded sites as an ornamental garden plant. It is sometimes called mace sedge because of its resemblance to the medieval weapon of that name.

DARK-GREEN BULRUSH
Scirpus atrovirens
PALE BULRUSH
Scirpus pallidus
SEDGE FAMILY *(Cyperaceae)*

Flowering period: May - Jun.

Occurrence: Near water along streams, lakes and marshes. **FF:** Uncommon along South Stream Trail. **NW:** Uncommon along MRE Trail.

Description: These two native perennials grow up to 5' tall on stout, triangular stems (culms). The leaves are up to 1" wide. Slender, leaf-like bracts are found just below the spherical clusters of spikelets. The clusters appear yellowish green in bloom, red-brown later on.

Identification: These two species are listed here together because they are very similar and not easy to tell apart. Pale bulrush *(S. pallidus)* has clusters up to 1/2" in diameter. Those of the dark-green bulrush *(S. atrovirens)* are usually about half that size.

RIVER BULRUSH
Scirpus fluviatilis; also Bolboschoenus fluviatilis.
SEDGE FAMILY *(Cyperaceae)*

Flowering Period: May - Jun.

Occurrence: Marshes, shorelines. **FF:** Rare near water off Marsh Trail. Abundant off the trails in water along the railroad tracks south of Handsome Hollow.

Description: This native perennial grows up to 5' tall. Its 1" wide, grooved leaves grow taller than the thick, triangular stem. Two or more bracts emerge from just below a bundle of short stalks. Each ends with a cluster of elongated, 1" long spikelets, which turn brown by the end of June.

REDROOTED SEDGE
Cyperus erythrorhizos
SEDGE FAMILY *(Cyperaceae)*

Flowering Period: August.

Occurrence: Muddy shores of streams and lakes. **FF:** Abundant along the shore of the Great Marsh near the blind.

Description: This native umbrella sedge has grooved leaves about as long as its main stem (culm). Flower spikes are arranged in compound clusters (umbels). Each spike has numerous tightly arranged, flat spikelets, which turn a rich copper-color at maturity.

Identification: Yellow nutsedge *(Cyperus esculentus)* has straw-colored spikelets. Fragrant sedge *(Cyperus odoratus)* has more loosely arranged straw-colored spikelets.

FRAGRANT SEDGE
Cyperus odoratus
SEDGE FAMILY *(Cyperaceae)*

Flowering period: Jul - Aug.

Occurrence: Lakeshores and wet, muddy soils. **FF:** Locally common in the field just east of the blind.

Description: This native umbrella sedge grows up to 30" tall in dense colonies. The flower spikes form compound umbels. Each spike has 4 to 30 straw-colored spikelets, densely crowded on its axis (rachis).

Identification: The similar yellow nutsedge *(Cyperus esculentus)* has spikelets more loosely arranged along its rachis. Red-rooted sedge *(Cyperus erythrorhizos)* has rich, copper-colored spikes, very densely clustered around its rachis.

YELLOW NUTSEDGE
Cyperus esculentus
SEDGE FAMILY *(Cyperaceae)*

Flowering period: July.

Occurrence: Muddy lake shores and ditches.
FF: Locally common in the open field near
the blind.

Description: This native, perennial umbrella
sedge grows from 6" to 30" tall. Its leaves
are about as long as the triangular stem
(culm). Several straw-colored flower spikes
form a compound umbel. Each spike has 8
to 20 loosely-arranged spikelets.

Identification: The very similar fragrant
sedge *(Cyperus odoratus)* has more densely
crowded spikelets. Red-rooted sedge
(Cyperus erythrorhizos) has copper-colored
spikelets, very densely clustered around its
axis (rachis).

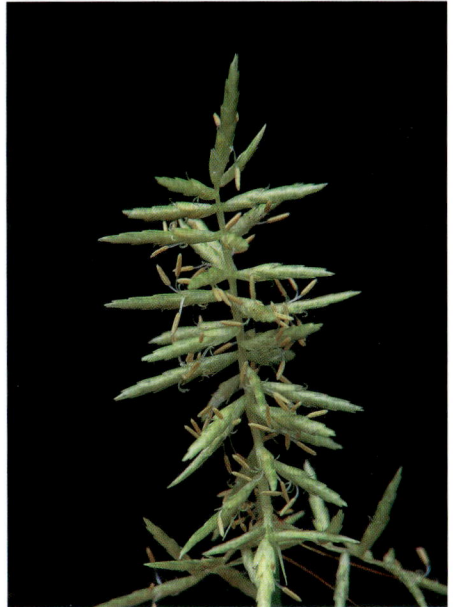

Comments: This nutsedge often presents a
serious weed problem in local lawns and
gardens.

PATH RUSH
Juncus tenuis
RUSH FAMILY *(Juncaceae)*

Flowering period: May - Jun.

Occurrence: Wet disturbed sites; along paths. **FF:** Abundant along woodland paths. **NW:** Abundant on path in Raccoon Hollow.

Description: This native plant grows in dense colonies, typically on trails or overhanging the sides of trails. Its stout stems reach up to 18" in height. Membranous auricles are present at the junction of the leaf and the stem (see close-up). Tiny white and green flowers, with 6 sharply pointed tepals, form a cyme. The fruit is a brown capsule, each holding many light brown seeds.

Identification: Very similar to Dudley rush *(Juncus dudleyi)*. The key feature distinguishing these two is the auricle, as shown in the close-ups.

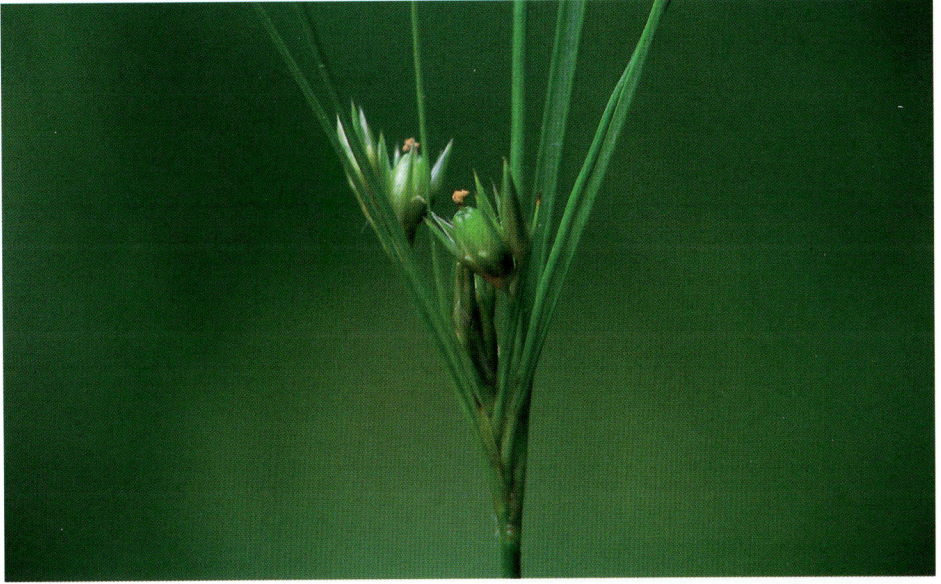

DUDLEY RUSH
Juncus dudleyi
RUSH FAMILY *(Juncaceae)*

Flowering Period: May - Jun.

Occurrence: Wet meadows and near water.
FF: Rare between Pond Trail and the pond.

Description: This native rush grows from 6"
to 36" tall, with stout stems. Its leaves are
half as tall as the stems. Rounded auricles
are present where the leaf joins the stem
(see close-up). The flower clusters form a
cyme; each flower has 6 sharply pointed
tepals. The brown fruit capsules hold many
light brown seeds.

Identification: Very similar to path rush
(Juncus tenuis). The key distinguishing
feature is the auricle (compare the close-
ups).

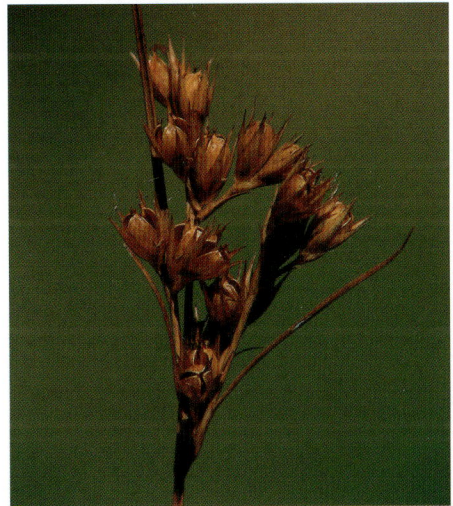

CAREX SPECIES LISTED BY SPIKE TYPE & HABITAT

SPIKES DIFFERENT
(MALE & FEMALE SPIKES SEPARATE)
Blunt-scaled Oak Sedge
Long-beaked Sedge
Woodland Sedge
Hitchcock's Sedge
Emory's Sedge
Shoreline Sedge (FF)
Smoothcone Sedge
James' Sedge*
Rocky Mountain Sedge*
Gray Wood Sedge
Few-fruited Sedge
Hairy Wood Sedge (NW)
Mead's Sedge (NW)
Davis' Sedge**
Bottlebrush Sedge (FF)
Bearded Sedge (FF)
Gray's Sedge (FF)

SPIKES ALL ALIKE
(MALES & FEMALES ON SAME SPIKE)
Sawbeak Sedge (FF)
Slender Sedge
Stellate Sedge
Heavy Sedge
Bur-reed Sedge
Woodbank Sedge
Soft Fox Sedge (FF)
Fox Sedge
Short-beaked Sedge
Troublesome Sedge

COMMENTS

*James' and Rocky Mountain sedges have very small male spikes and a row of perigynia below - both have very prominent bracts partially or completely enclosing the flower cluster unlike any of our other sedges.

**A number of sedges with different spikes may have one or more of the upper spikes with both male and female flowers. In this case the male flowers are always on the top. Our only exception is Davis' sedge which has female flowers above the male flowers on the upper spike.

Sedges with (FF) or (NW) after the name have been found only at that site.

UPLANDS
SPIKES DIFFERENT
(MALE & FEMALE SPIKES SEPARATE)
Blunt-scaled Oak Sedge
Long-beaked Sedge
Woodland Sedge
Hitchcock's Sedge
James' Sedge*
Rocky Mountain Sedge*
Gray Wood Sedge
Few-fruited Sedge
Hairy Wood Sedge (NW)
Davis' Sedge**

SPIKES ALL ALIKE
(MALES & FEMALES ON SAME SPIKE)
Stellate Sedge
Heavy Sedge
Bur-reed Sedge
Woodbank Sedge
Short-beaked Sedge
Troublesome Sedge

FLOODPLAIN
SPIKES DIFFERENT
(MALE & FEMALE SPIKES SEPARATE)

Found in water or very moist sites
Emory's Sedge
Shoreline Sedge (FF)
Smoothcone Sedge
Bottlebrush Sedge (FF)
Bearded Sedge (FF)

Found in drier sites
Long-beaked Sedge
Woodland Sedge
James' Sedge*
Gray Wood Sedge
Few-fruited Sedge
Davis' Sedge**
Gray's Sedge (FF)

SPIKES ALL ALIKE
(MALES & FEMALES ON SAME SPIKE)

Found in water or very moist sites
Sawbeak Sedge (FF)
Fox Sedge

Found in drier sites
Slender Sedge
Stellate Sedge
Heavy Sedge
Soft Fox Sedge (FF)
Short-beaked Sedge
Troublesome Sedge

Grasses

This third part features 60 species descriptions of grasses found at FF/NW. Only one species is included on each page except in a few cases, mainly the *Muhlenbergias,* where individual species cannot be reliably separated by the non-technical characters used in this book.

Superficially, grasses share many features with other grass-like plants, the sedges and rushes. The characters in this paragraph that are highlighted in bold separate the grasses from the sedges. The culm or flowering **stem** is usually **round and hollow** and is interrupted at intervals by swollen areas called **nodes (joints).** Leaves have two parts: the blade and a less obvious portion, the sheath, which wraps around the stem and extends down to the node below. Typically, all or a portion of the **leaf sheath is split,** a feature referred to as an open or partially open sheath (see Illustrated Glossary).

Grasses and rushes share many of the above features including the round stem; **unlike the rushes, grasses have jointed, hollow stems.**

A basic understanding of the components of the flower cluster or inflorescence is important for grass identification. Important features are highlighted in bold below and illustrated in the glossary at the back of this book. The flower cluster is made up of a number of subdivisions called spikelets. A typical **spikelet** has two modified bracts called **glumes** at its base. Above the glumes are one to many **florets or flowers.** Each floret has two additional bracts called the **palea** and **lemma** which enclose the reproductive parts, the anthers (usually 3) and the two feathery stigmas. They often protrude from their enclosing bracts (palea and lemma) and may be visible when the grass is flowering. Glumes, lemmas or paleas may have bristle-like extensions called **awns.**

Two other characters often used as **identifiers** are the auricle and ligule. **Auricles** are a pair of ear-like extensions of the blade margins which wrap around the stem (see Illustrated Glossary). The **ligule** is a membranous flap or a row of hairs on the inside of the upper sheath. Since the ligule usually can be seen only when the sheath is peeled back from the stem, it will rarely be used as an identifier in this guide.

Tips on using this portion of the book

- Grasses are arranged roughly according to their first flowering dates, except in cases where similar species are grouped together on facing pages so they may be more easily compared.
- The habitat in which the plant occurs at FF/NW and unique characters that identify it are underlined.
- In cases where a more detailed comparison with a similar species is indicated and space permits, an identification section is included.
- A table dividing grasses into habitat groupings is included at the end of this section. This table is designed to narrow the number of species choices from 60 to 25 or less.
- A hand lens will be necessary to identify the smaller grass features described in this book (or, perhaps, to decipher the fine print in the table)!

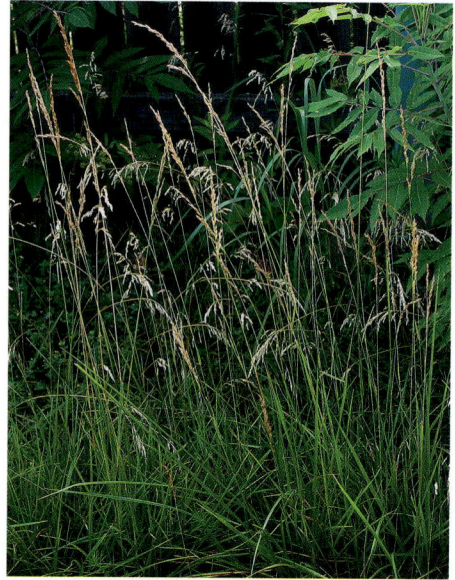

KENTUCKY BLUEGRASS
Poa pratensis
GRASS FAMILY *(Poaceae)*

Flowering Period: May - Aug; mostly May - Jun.

Occurrence: Widely distributed in a variety of habitats. **FF:** Common along Hidden Lake Trail. **NW:** Common along MRE Trail.

Description: Strongly rhizomatous sod-forming perennial with erect slender stems 4-40" tall. The numerous finely textured basal leaves have rather abruptly tapered tips resembling the prow of a boat and measure from 2-10" long by less than 1/4" wide. Flower clusters are pyramidal whorled panicles with 3-5 branches in each whorl (lower right). Spikelets, each containing 3-6 flowers, are at the tips of the panicle branches. In flower the panicles are open, but contract after flowering (upper right).

Identification: See discussion under woodland bluegrass *(Poa sylvestris).*

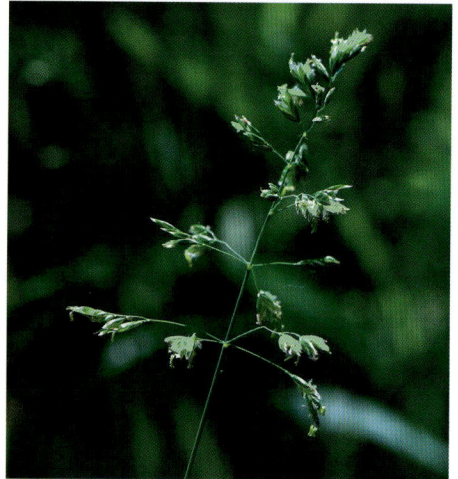

Comments: Kentucky bluegrass may be a North American native, but most plants are believed to be of European origin. Thriving in many habitats, it has one of the broadest distributions of any temperate zone plant. A good forage grass, it withstands grazing, often replacing native species in overgrazed prairies. Its color, texture, hardiness and turf-forming habit have made it a very popular lawn grass.

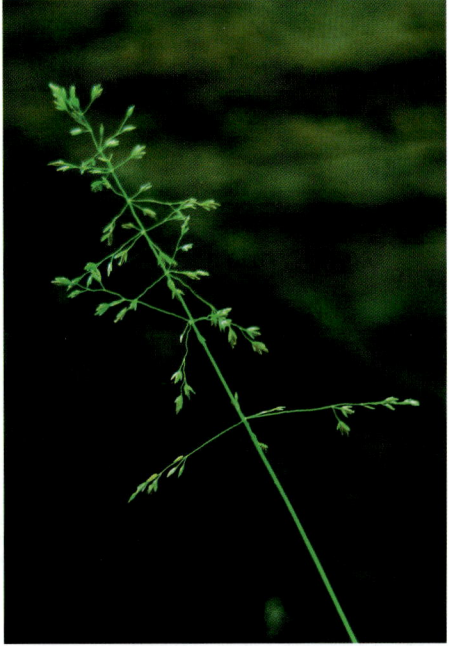

WOODLAND BLUEGRASS
Poa sylvestris
GRASS FAMILY *(Poaceae)*

Flowering Period: May - Jun.

Occurrence: <u>Moist woodlands</u>. **FF:** Uncommon in Childs Hollow. **NW:** Uncommon in Settlers Hollow.

Description: Native, <u>bunched perennial</u> with erect stems up to 36" tall. Flat <u>leaf blades</u> with boat-shaped tips are 2-6" long and <u>up to 1/4" wide</u>. Flower cluster is an open panicle most often consisting of <u>sets of 5-6 spreading, often drooping branches arranged in whorls at the lower nodes</u>. Spikelets, each with 2-5 florets, occur near the tips of the panicle branches.

Identification: Unlike woodland bluegrass, the significantly more common Kentucky bluegrass *(Poa pratensis)* is a sod-forming grass which forms extensive rhizomes. The lower whorl of the Kentucky bluegrass panicle usually has fewer (4-5) branches which are spreading or ascending, not drooping. Other more subtle characters overlap, but Kentucky bluegrass leaves generally are narrower and the panicle slightly smaller and less open than woodland bluegrass.

ANNUAL BLUEGRASS
Poa annua
GRASS FAMILY *(Poaceae)*

Flowering Period: Apr - Oct; mostly May - Jun.

Occurrence: <u>Moist areas along trails,</u> roadsides, ditches, lawns, gardens, waste areas. **FF:** Common along trails in Childs Hollow. **NW:** Common on trail in Raccoon Hollow.

Description: Short and inconspicuous introduced weedy annual from <u>2-10" tall</u> growing in bunches, sometimes forming large mats. The flat, narrow blades are up to 5" long and no more than 1/8" wide. Flowering stems may be erect or, more often, spreading. Each bears an <u>open panicle</u> from 1-6" long which has rather <u>prominent spikelets</u> containing from 2-6 florets at the tips of the <u>solitary or paired panicle branches</u>.

Identification: None of the other short grasses under 1' tall growing along or on our paths have open panicles with prominent spikelets like those of annual bluegrass.

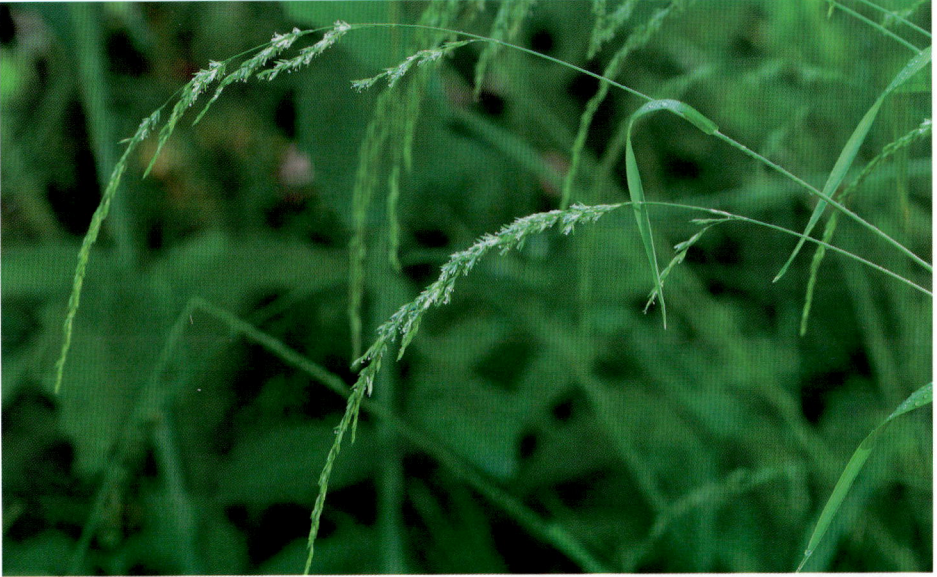

SLENDER WEDGEGRASS
Sphenopholis intermedia; also
Sphenopholis obtusata var. *major*
GRASS FAMILY *(Poaceae)*

Flowering Period: May - Jun.

Occurrence: <u>Moist ground usually in the woods.</u> **FF:** Locally common near North Stream/Cottonwood Trail junction. **NW:** Uncommon on River Trail.

Description: Native annual or perennial growing as a solitary stem or more often in bunches. The <u>thin</u> 1-3' <u>stems</u> have <u>rather inconspicuous leaves</u> from 2-8" long and up to 1/4" wide. The flower cluster is a moderately open to densely closed, <u>slender, nodding 2-8" panicle.</u>

Identification: Nodding fescue *(Festuca subverticillata)* and fowl mannagrass *(Glyceria striata)* occupy similar habitats and have nodding panicles. Panicles of both are significantly more open and spreading, and they have larger, more prominent leaves.

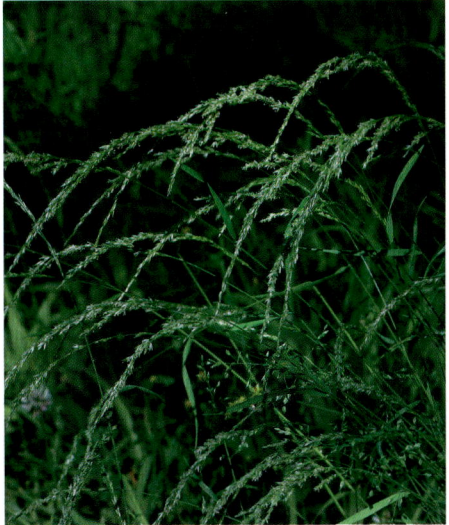

Comments: This inconspicuous grass matures quickly during the month of June and disappears much like our early woodland wildflowers.

NODDING FESCUE
Festuca subverticillata; also
Festuca obtusa
GRASS FAMILY *(Poaceae)*

Flowering Period: Late May - Jul.

Occurrence: <u>Moist upland and lowland woods</u>. **FF:** Common along GM boardwalk. **NW:** Common along MRE Trail.

Description: Native perennial with stems from 1 1/2 to 3' tall. Plants may be solitary but often grow in small bunches. Leaves are up to 12" long and less than 1/2" wide. Flower clusters are <u>large, open, spreading panicles</u> up to 12" long which usually <u>droop noticeably at maturity</u>. From 2-7 <u>spikelets,</u> each consisting of 2-5 individual florets, are present <u>at the tips of the panicle branches</u> (lower).

Identification: Fowl mannagrass *(Glyceria striata)* has similar large, open panicles, but spikelets are more evenly distributed along the branches, and it is usually found in wetter habitats. Slender wedgegrass *(Sphenopholis intermedia)* panicles are more compact, often densely closed. Whitegrass *(Leersia virginica)* flowers later, and its panicles do not droop.

FOWL MANNAGRASS
Glyceria striata
GRASS FAMILY *(Poaceae)*

Flowering Period: Late May - Jul.

Occurrence: <u>Wet woodlands, wet ground along streams</u> and ponds, freshwater wetlands. **FF:** Uncommon in Child's Hollow. **NW:** Uncommon along MRE Trail near Rock Creek.

Description: Rhizomatous native perennial with erect to leaning flowering stems up to 4' tall. Leaf blades are narrow, usually no more than 1/4" wide. The flower cluster is an <u>open, nodding panicle</u> up to 10" long bearing a <u>large number of spikelets which extend nearly all the way to the base of the panicle branches.</u>

Identification: Nodding fescue *(Festuca subverticillata)* is usually found in drier sites, the panicle is more open, and spikelets are confined to the tips of the panicle branches. The panicle of slender wedgegrass *(Sphenopholis intermedia)* is dense and much narrower.

BUFFALO GRASS
Buchloe dactyloides
GRASS FAMILY *(Poaceae)*

Flowering Period: May - Jun.

Occurrence: <u>Prairies</u>. Common and often dominant in short grass prairies to the west. Found in our area mostly on drier exposed sites. **FF:** Rare along Hidden Lake Trail. **NW:** Uncommon on upper Gifford Trail (Jonas Prairie) beyond the bench.

Description: <u>Very short</u> native perennial up to 8" tall <u>often forming dense mats</u>. This unusual species has above-ground stolons or runners and occurs as separate male and female plants (male, middle; female, lower photo). The inconspicuous bur-like female flowers are on short stems near the ground, closely associated with modified leaves. The more visible <u>comb-like male flower spikes</u> occur on slender stems that rise above the leaves, remaining in place well after flowering.

Comments: Buffalo grass is an important range grass utilized by all grazers. Its drought resistance, fine texture and short stature have prompted use as a turf grass.

SCRIBNER'S PANICGRASS

Panicum oligosanthes var. *scribneri-anum:* also *Dichanthelium oligosanthes* var. *scribnerianum*
GRASS FAMILY *(Poaceae)*

Flowering Period: Apr - Jun. Secondary panicles blooming until Sep.

Occurrence: <u>Prairies</u>, disturbed areas, occasionally in woods. **NW:** Uncommon in Koley Prairie.

Description: Rather <u>short and inconspicuous</u> bunched native perennial from 4-28" tall. Photos (above and right) show spring flower clusters which are pyramid-shaped <u>panicles with wavy branches</u> terminating in small <u>egg-shaped spikelets</u>, often with a reddish spot at the base. Flower clusters occurring later in the season are much smaller with fewer branches. Leaves are arranged in a rosette with shorter, wider ones at the base

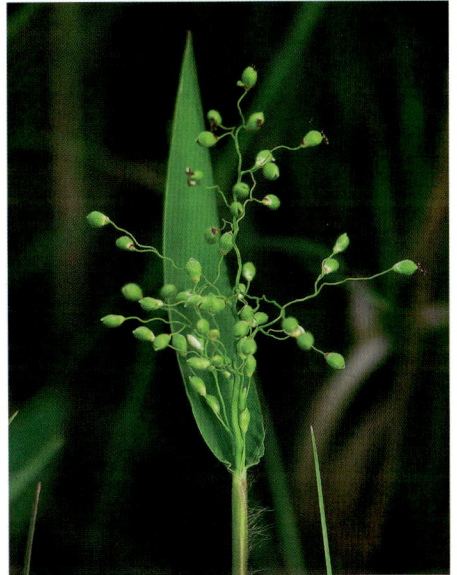

and longer, narrower leaves on upper stems which become more branched as the season progresses. Rosettes form in early fall and remain green all winter.

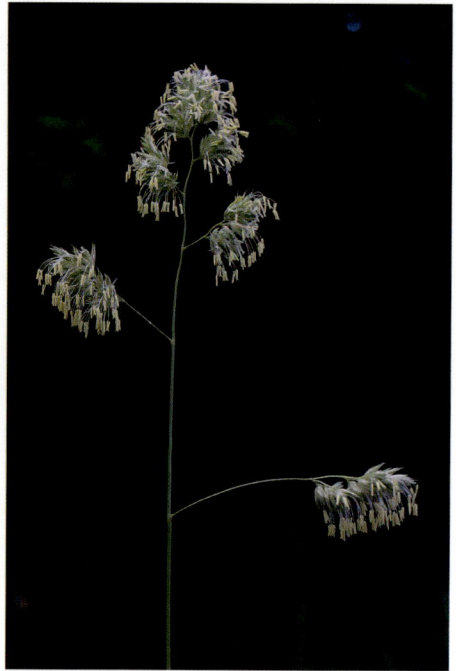

ORCHARD GRASS
Dactylis glomerata
GRASS FAMILY *(Poaceae)*

Flowering Period: Mostly May - Jun; occasionally as late as Oct.

Occurrence: Disturbed areas in many habitats including fencerows, roadsides, pastures, woodlands or prairies. **FF:** Common on Marsh Trail and GM Boardwalk. **NW:** Uncommon on Nebraska Trail at edge of prairie.

Description: Introduced perennial bunchgrass often forming large clumps. Erect flowering stems 2-4' tall bear compact to open panicles with only a few major branches, the longest ones at the bottom. These branches have dense, rounded-to-oval clusters of spikelets near their ends, the spikelets oriented predominantly along one side of the axis. Arching leaves are 4-18" long and less than 1/2" wide.

Identification: No other tall cool season grass at FF/NW has all spikelets on one side of the branch.

Comments: Orchard grass was introduced from Europe in the 1700's for hay and forage. It is now naturalized where winters are not too severe and rainfall is adequate.

REED CANARY GRASS
Phalaris arundinacea
GRASS FAMILY *(Poaceae)*

Flowering Period: Late May - Jul.

Occurrence: <u>Wet ground</u> in low areas, marshes and ditches. **FF:** Common along North and South Stream Trails. **NW:** Common on portions of MRE Trail.

Description: Strongly rhizomatous sod-forming perennial forming <u>dense, almost impenetrable colonies in open moist sites</u>. The broad (up to 1" wide), smooth leaf blade has a prominent, <u>usually visible, white membranous ligule</u> at its base (lower right). Flowering stems from 3-6' tall bear <u>compact lobed panicles with ascending branches</u>. In flower, the panicles are open and often tinged with red or purple (middle right). They contract and fade to a light straw color at maturity (upper right).

Comments: This variable species, which includes both native and introduced populations, has been planted for hay, forage and erosion control. Once established, it can be difficult to control. It is the dominant plant and spreading in many moist, open floodplain sites at FF/NW.

143

SMOOTH BROME
Bromus inermis
GRASS FAMILY *(Poaceae)*

Flowering Period: Late May - Jul.

Occurrence: <u>Fields, pastures, roadsides, ditches.</u> **FF:** Locally common on Ridge and North Stream Trails. **NW:** Abundant in open fields along Jonas and Hilltop Trails.

Description: Strongly rhizomatous, sod forming, introduced perennial with flowering stems from 1 1/2 - 3 1/2' tall. The <u>stem, leaves and leaf sheaths are smooth and hairless.</u> Although present in other bromes, the <u>M-shaped constriction of the leaf</u> (lower), is especially prominent in smooth brome and a good field mark. The moderately open symmetrical 5-8" panicles have ascending branches bearing <u>large, straw-colored spikelets from 3/4 - 1 1/4" long</u>, each containing several florets.

Identification: Our other brome species, ear-leaf *(B. latiglumis),* Canada *(B. pubescens)* and Japanese brome *(B. japonicus)* have nodding panicles and hairy leaf sheaths. The first two are found in shadier sites.

Comments: This cool season grass was introduced from Eurasia in the 1880's and is still widely cultivated for cover, pasture and hay. It has escaped into a variety of habitats throughout much of the U.S. and Canada.

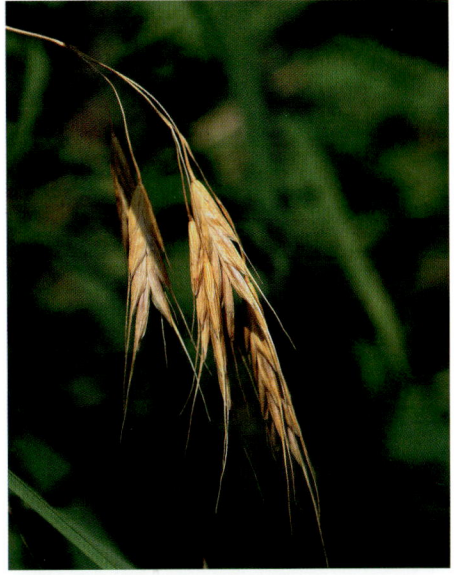

JAPANESE BROME
Bromus japonicus
GRASS FAMILY *(Poaceae)*

Flowering Period: Late May - Jul.

Occurrence: <u>Disturbed areas;</u> waste ground.
FF: Locally common in floodplain prairie planting just beyond the blind. **NW:** Rare along lower Gifford Trail at edge of woods.

Description: Introduced annual with upright or ascending stems up to 24" tall and narrow, hairy leaves. The <u>open, nodding,</u> often wavy <u>panicle branches</u> bear numerous <u>hairless spikelets</u> composed of 6-13 florets, each tipped with an awn up to 1/2" long that is bent or spread outward at maturity. <u>Leaf sheaths are densely covered with long, soft white hairs</u>.

Identification: Dense stands of very similar downy brome *(Bromus tectorum)* occur occasionally at RR track crossings, but have not been seen elsewhere in FF. Unlike Japanese brome, its spikelets are often reddish-purple at maturity and are covered with fine hairs, hence the name 'downy'.

TALL FESCUE
Lolium arundinaceum; also
Festuca arundinacea
GRASS FAMILY *(Poaceae)*

Other Common Names: Kentucky-31, Alta, reed and coarse fescue.

Flowering Period: Mostly May - Jul; occasionally to Oct.

Occurrence: <u>Disturbed sites, roadsides,</u> ditches, pastures. **FF:** Common along GM boardwalk, especially in planting near WLC. **NW:** Common and planted in River Road ditch in front of ranger's residence and on the dam in Raccoon Hollow.

Description: Introduced rhizomatous perennial with stout, erect stems from 2-4 1/2' tall. Dark green leaves are flat, long and narrow with pointed tips. At the junction of the leaf and its smooth sheath are a pair of <u>ear-like appendages or auricles.</u> The flower cluster is an <u>asymmetrical narrow-to-open 6-10" long panicle with ascending branches</u> bearing 5-15 spikelets at the tip.

Identification: Although the majority of our fescues are believed to be tall fescue, another introduced species, meadow fescue *(Lolium pratense),* is also present. Subtle differences in the auricles separate them. Smooth brome *(Bromus inermis)* has a more symmetrical panicle, larger cylindrical spikelets and no auricles.

Comments: A native of Europe, tall fescue has been planted extensively for use as hay, forage and erosion control. Low-growing races are also commonly used as a turf grass. A mutualistic fungus often infects this plant, producing a toxin which affects a small percentage of livestock grazing on infected fescue. They develop a condition known as "summer slump" characterized by reduced weight gain, reduced milk production and heat stress due to increased body temperature. It also can cause lameness known as "fescue foot".

CANADA BLUEGRASS
Poa compressa
GRASS FAMILY *(Poaceae)*

Flowering Period: Jun - Aug.

Occurrence: <u>Wide variety of habitats</u>, most often in disturbed sites. **FF:** Rare along Stream and Missouri Trails.

Description: Strongly rhizomatous perennial often growing in clumps or loose colonies. Wiry <u>stems are flattened</u>, especially in their upper portions as shown in the front, side and cross-sectional views (upper and lower right). Stems are up to 24" tall and bear a fairly dense, compact or slightly open panicle which usually has no more than 2 branches at each node. The ascending leaves are short (up to 4" long), narrow, and shaped like the prow of a boat at the tip.

Identification: The flattened stem separates Canada bluegrass from our other bluegrass species which have round, or nearly round stems. This feature makes it difficult to roll the stem when grasped between the thumb and index finger.

Comments: Introduced from Eurasia in the late 1700's, Canada bluegrass competes poorly with Kentucky bluegrass *(Poa pratensis)* in good soils. Canada bluegrass does better in poorer soils or drier situations and is sometimes used for cover or erosion control in disturbed sites.

LITTLE BARLEY
Hordeum pusillum
GRASS FAMILY *(Poaceae)*

Flowering Period: Late May - Jun.

Occurrence: <u>Disturbed ground</u>. **FF:** Locally common in recent floodplain prairie planting east of the blind.

Description: Native annual usually growing in bunches. The erect 4-16" flowering stems curve abruptly upward from a short basal horizontal segment. Narrow, short leaf blades up to 4 1/2" long are ridged on the upper surface. The <u>erect to slightly curved, spike-like flower cluster</u> (technically a raceme) is from 1 1/2 to 3" long. It is often

partially enclosed in the upper leaf sheath and consists of many clusters of 3 spikelets, only the central one fertile. All spikelets, however, have glumes and lemmas with short awns responsible for the <u>bristly appearance</u>. At maturity the flower cluster breaks apart from the top down (lower).

FOXTAIL BARLEY
Hordeum jubatum
GRASS FAMILY *(Poaceae)*

Flowering Period: Jun - Aug.

Other Common Name: Squirrel tail.

Occurrence: Waste ground, pastures, ditches and roadsides. **FF:** Uncommon on floodplain along edge of Camp Gifford Road.

Description: Short-lived, bunched native perennial 12-32" tall. Flat leaf blades are up to 5" long and 1/4" wide tapering to a pointed tip. The drooping 3-5" flower spike with its straight, spreading, long awns responsible for its characteristic "bushy" appearance makes identification of this grass easier than most. The soft, greenish or purplish spike fades to a pale brown at maturity when the spike breaks apart.

Identification: A related species, little barley *(H. pusillum)*, has erect seed heads with short awns.

Comments: The long awns can injure grazing livestock or wildlife, producing sores in the mouth, throat, nose and eyes. They also lodge in fleece, thereby contaminating the wool. This native grass was collected by Lewis & Clark in Montana and near Fort Clatsop in 1806.

WESTERN WHEATGRASS
Elymus smithii; also *Pascopyrum smithii; Agropyron smithii*
GRASS FAMILY *(Poaceae)*

Flowering Period: June - Aug.

Occurrence: <u>Prairies, roadsides</u>, ditches, waste areas. **FF:** Uncommon along GM boardwalk near WLC. **NW:** Uncommon in lower Jonas Prairie along Gifford Trail.

Description: Strongly rhizomatous native perennial with stems from 1-3' tall. The waxy <u>stem and leaves have a distinctive bluish coloration</u>, and the stiff, narrow leaf blades are usually rolled inward on the edges. <u>Prominent ridged veins on the upper surface</u> (lower right) feel rough to the touch. Auricles are usually prominent. The flower cluster is an erect 2-8" spike consisting of a <u>double row of alternating, overlapping spikelets</u> (lower left).

Identification: Quackgrass *(Elymus repens)* has a similar spike with overlapping spikelets, but its leaves are flat, dark green and not prominently ridged. The spike of tall wheatgrass *(Elymus elongatus)* is open with more widely separated spikelets.

QUACKGRASS
Elymus repens; also
Agropyron repens
GRASS FAMILY *(Poaceae)*

Flowering Period: Jun - Aug.

Occurrence: <u>Moist areas</u> including streambanks, pastures, gardens, lawns, ditches and cultivated fields. **FF:** Rare in vegetation between WLC and Camp Gifford Road. **NW:** Uncommon and local along Missouri River bank near MRE Trail benches.

Description: Introduced, strongly rhizomatous perennial, a native of Europe, with flowering stems up to 3 1/2' tall. <u>Flat blades</u> up to 1/2" wide and 12" long are <u>not prominently ridged</u> (lower right), but usually have prominent auricles at their bases. Flower heads are <u>spikes with 2 rows of alternate, overlapping spikelets</u> oriented along the long axis of the stem (lower left). Glumes and lemmas usually have short awns.

Identification: Western wheatgrass *(Elymus smithii)* has a similar spike, but leaves are prominently ridged and usually rolled up on the edges. Spikes of tall wheatgrass *(Elymus elongatus)* are open, with the spikelets widely separated.

Comments: Quackgrass is a good hay and forage plant, but its aggressive habit of invading moist grazing land and cultivated fields makes it a serious weed.

HAIRY WILD RYE
Elymus villosus
GRASS FAMILY *(Poaceae)*

Other Common Names: Downy wild rye; slender wild rye.

Flowering Period: Late May - Jul.

Occurrence: Dry to moist upland or lowland woods and edges. **FF:** Common along Hawthorn Trail in savanna restoration. **NW:** Common on upper Neale Trail.

Description: Native perennial growing singly or in bunches. Slender stems are 1 1/2 to 4' tall. Shiny, 2-7" long, dark green leaves have soft, velvety hairs on the upper surface. The 2-4" long spikes, straight at first but characteristically arching at maturity, extend well above the leaves. The narrow, straight glumes, responsible for the bristly appearance of the spike, persist after the seeds fall, often until the following spring.

Identification: Virginia wild rye *(Elymus virginicus)* has straight spikes that usually are partly enclosed in the upper leaves. Glumes are broader and fall with the seeds leaving a naked stem. Canada wild rye *(Elymus canadensis)* has longer, thicker arching spikes with long awns that are curved and twisted at maturity giving the spike a shaggy look.

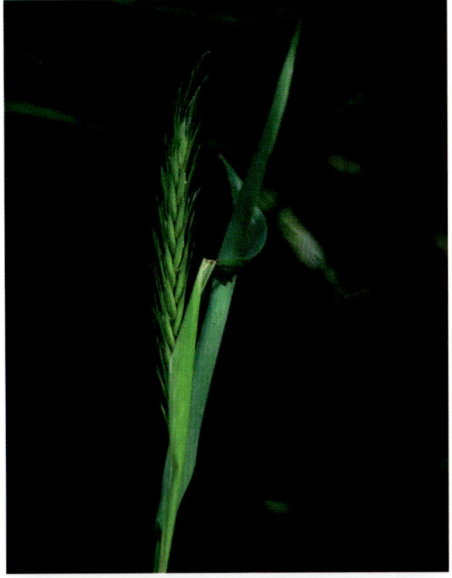

VIRGINIA WILD RYE
Elymus virginicus
GRASS FAMILY *(Poaceae)*

Flowering Period: Late May - Jul.

Occurrence: <u>Floodplain woodlands,</u> streambanks, low prairies, woodland edges. **FF:** Common along Marsh Trail. **NW:** Common on River Trail.

Description: Native perennial growing singly or in bunches, sometimes forming large colonies on the floodplain. Flowering stems of this plant, which is quite variable in size and growth habit, are from 1-5' tall. The stiff and <u>straight flowering spikes</u> are usually <u>partially enclosed by the upper leaf sheath</u> and are 1 1/2 to 6" long. Characteristic <u>straplike, U-shaped glumes</u> (lower left) provide a good field mark. When mature, the glumes and enclosed seeds fall together leaving a naked stem (lower right).

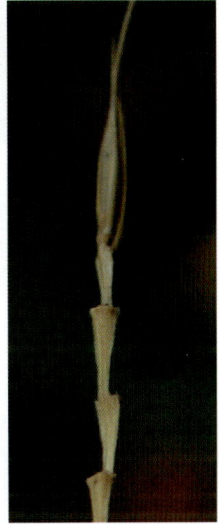

Identification: Hairy wild rye *(Elymus villosus)* has nodding spikes projecting well above the leaves and narrow, straight glumes which are retained on the stem when

the seeds fall. Canada wild rye *(Elymus canadensis)* has larger nodding spikes, and the florets have long curved and twisted awns giving the spike a "shaggy" appearance.

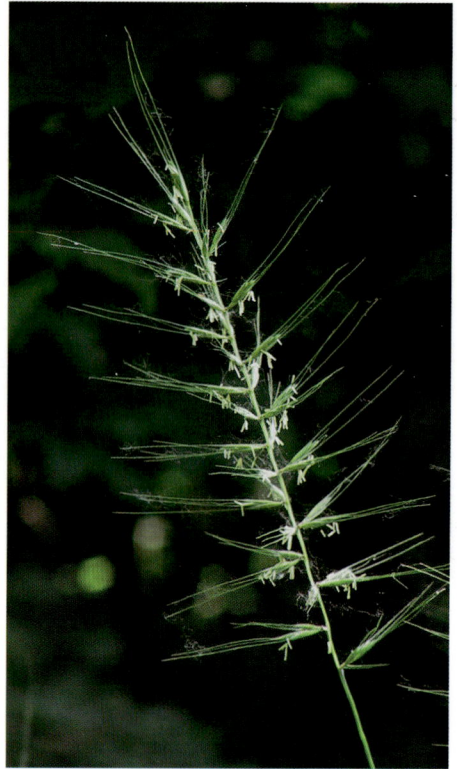

BOTTLEBRUSH GRASS
Elymus hystrix; also *Hystrix patula*
GRASS FAMILY *(Poaceae)*

Flowering Period: Jun - Aug.

Occurrence: Rich soil in upland or lowland woods. **FF:** Common on floodplain along Cottonwood Trail. Good upland display along Hawthorn Trail in savanna restoration. **NW:** Common on MRE Trail near bridge.

Description: Native perennial with long slender stems from 2-4' tall, growing singly or in small bunches. Evenly spaced leaves up to 1/2" wide are overshadowed by the showy, open 4-6" flower spikes. Paired spikelets, each with 2-4 florets bearing awns up to 1 1/2" long, come off the central stem at right angles creating the "bottlebrush-like"

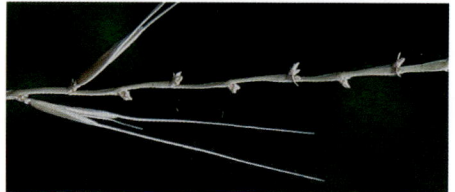

flower cluster responsible for its common name. The mature spikelets drop, leaving a "naked" stem (lower right) which often persists through the winter.

Identification: The distinctive flower cluster makes this, perhaps, our easiest grass to identify. After the spikelets have dropped, the stem looks very much like Virginia wild rye *(E. virginicus)*. Close examination reveals it is more slender, with knobbier projections where the spikelets were previously attached.

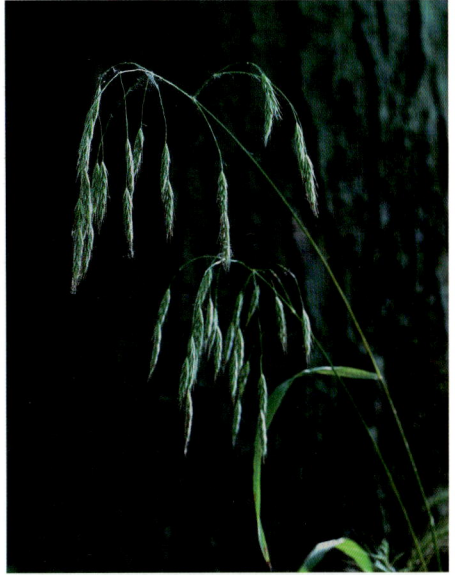

CANADA BROME
Bromus pubescens
GRASS FAMILY *(Poaceae)*

Flowering Period: Jun - early Jul.

Occurrence: Upland and floodplain woods.
FF: Common on Cottonwood and North Stream Trails. **NW:** Uncommon on Neale and Columbine trails.

Description: Slender, bunched native perennial with upright or leaning stems 2 1/2 - 5' tall. Leaves are often softly hairy, especially above. There are no more than 4-6 leaves per stem. Soft, thin hairs are present on the leaf sheaths and stems (lower photos). Stem hairs are most dense about the nodes, which are visible and not obscured by the leaf sheaths (lower right). No ear-like appendages, or auricles, are present at the leaf base in this species. The flower cluster is an open, drooping 4-10" panicle bearing spikelets up to 1" long, each with 4-8 florets.

Identification: Another woodland brome with drooping panicles, ear-leaf brome *(B. latiglumis),* flowers in August and

September, well after Canada brome has flowered. Ear-leaf brome also is leafier with more than 10 leaves per stem and has auricles at the base of the leaf blade. Smooth brome *(B. inermis),* found in more open habitats, has erect panicles and hairless stems and leaves. Japanese brome *(B. japonicus)* has nodding panicles and hairy leaf sheaths, but is shorter (less than 2') and occupies open, disturbed sites.

EASTERN GAMA GRASS
Tripsacum dactyloides
GRASS FAMILY *(Poaceae)*

Flowering Period: Jun - Sep.

Occurrence: <u>Moist prairies</u>, bottomlands, riverbanks. **FF:** Locally common on dam in Child's Hollow.

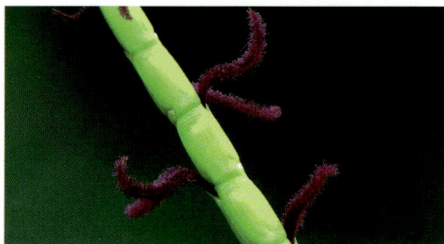

Description: Native perennial forming <u>large clumps from 3-8' tall</u>. At the tip of a stout, long stalk are 1-4 flower spikes. <u>Male and female flowers are separate,</u> the lower portion of the spike consisting of a series of cylindrical segments, each containing a single female spikelet with large, purple stigmas (middle). The slender upper portions have two rows of male spikelets aligned along one side of a flattened stem.

Comments: Although native to SE Nebraska, our plants originated from seed mix used to cover the Childs Hollow dam. Eastern gama grass is closely related to corn *(Zea mays).*

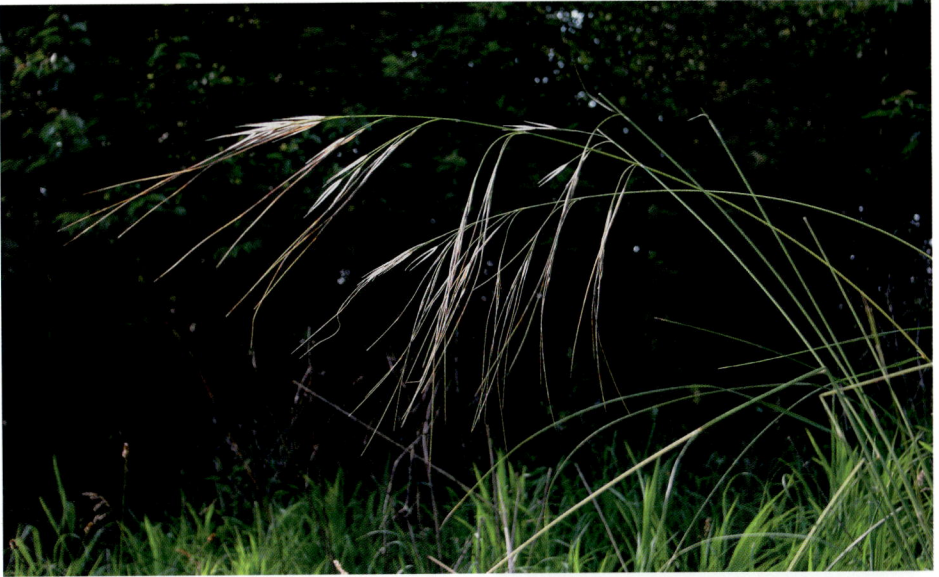

PORCUPINE GRASS
Hesperostipa spartea; also
Stipa spartea
GRASS FAMILY *(Poaceae)*

Flowering Period: Jun - Jul.

Occurrence: <u>Upland prairies</u>. **NW:** Rare in Nebraska Prairie.

Description: Native perennial with flowering stems from 1 1/2 to 4' tall. Long, narrow leaves with rolled-up margins are up to 18" long and less than 1/8" wide. The mature <u>arching flower stems</u> bear <u>long, narrow panicles with drooping clusters of flowers</u> at the branch tips. Each flower and, eventually, the maturing seed is enclosed by two pale bracts called glumes. (Lower photo shows empty glumes from which the seed has fallen just above those still enclosing the narrow, 1" long seed.) Two additional bracts cover the seed. One, the reddish-brown lemma visible in the photo, has a <u>characteristic 4-7" sharp-pointed, twisted and coiled bristle-like awn</u> at its tip.

Comments: The sharp seed tips and long awns can cause injury to grazing livestock and are responsible for names like needlegrass and needle-and-thread for other similar members of this genus. The awns are remarkable self-planting devices, actually drilling into the soil as they coil and uncoil in response to changing moisture conditions.

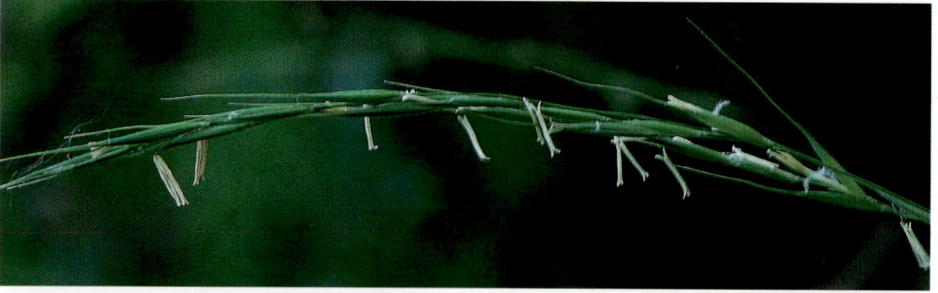

BEARDED WOOD GRASS
Brachyelytrum erectum
GRASS FAMILY *(Poaceae)*

Flowering Period: Late Jun - Aug.

Occurrence: Moist upland woods. **FF:** Common on Prairie Trail.

Description: Native perennial with erect or nodding flower stems from 2-3' tall. The flat, relatively short 3-8" leaves are broadest near the base, tapering slowly to a bluntly pointed tip. Most leaves come off one side of the stem and project out stiffly at a 90 degree angle, giving individual plants a "windblown" look. Cross-veins can be seen between the main parallel veins when transilluminated or backlit by the sun (see lower right photos without and with magnification). The flower cluster is a narrow, closed panicle from 3-6" long bearing slender 1/2" florets each with an awn that is usually longer than the floret (upper).

Identification: Another superficially similar plant of the upland woods, black-seed ricegrass *(Piptatherum racemosum),* has an open panicle and longer arching leaves which do not have cross veins. The cross-veins are unique and not seen in any other grass at FF/NW.

BLACK-SEED RICEGRASS
Piptatherum racemosum; also
Oryzopsis racemosa
GRASS FAMILY *(Poaceae)*

Flowering Period: Late Jun - Aug.

Occurrence: <u>Upland woods</u>. **FF:** Uncommon on Oak Trail. **NW:** Rare; found in only one site on west Maidenhair Trail.

Description: Native perennial growing in small bunches. Thin, erect-to-arching stems 20-36" tall bear <u>open sparsely branched panicles</u>. Lower leaves are quite small, but upper ones are better developed, measuring 6-12" long and up to 1/2" wide. Leaf blades are twisted 180 degrees at the base causing the true upper side of the leaf to rest underneath the bottom side (lower right). The mature <u>rice-like seeds</u>, each bearing a long awn, <u>are aligned along the long axis of the panicle branches</u> resembling a string of elongated pearls (upper right).

Identification: Whitegrass *(Leersia virginica)* has a similar open panicle. Unlike

black-seed ricegrass, it is a reclining plant with weak stems and white hairs on the nodes.

Comments: The common name is derived from the dark lemma covering the seed. It is concealed by the glumes which have been pulled back to expose it in the photo (lower left).

AMERICAN BEAKGRAIN
Diarrhena obovata; also
Diarrhena americana var. *obovata*
GRASS FAMILY *(Poaceae)*

Other Common Names: Wood grass, forest grass, obovate beakgrain.

Flowering Period: Late Jun - Sep.

Occurrence: Upland or lowland woods. **FF:** Common on Cottonwood and Prairie Trails. **NW:** Common on Deer and Hilltop Trails.

Description: Native rhizomatous perennial with flowering stems 1-4' tall growing singly or more often in loosely grouped colonies, some quite extensive. In spring and early summer numerous broad, glossy, dark green basal leaves up to 3/4" wide are the most prominent feature of this attractive woodland grass. In late June arching flower stems bearing long, closed panicles made up of many very ordinary-appearing 2-5 flowered spikelets appear (middle). By late summer

the mature florets develop into distinctive pear-shaped grains with prominent beaked tips, unlike any of our other grasses (lower).

WHITEGRASS
Leersia virginica
GRASS FAMILY *(Poaceae)*

Flowering Period: Jul - Oct.

Occurrence: <u>Moist upland and floodplain woods</u>. **FF:** Common in Handsome Hollow. **NW:** Common on woodland portions of Jonas Trail and in Raccoon Hollow.

Description: Rhizomatous perennial 12-18" tall with long, <u>slender, weak, spreading stems</u> often forming large sprawling patches. <u>Nodes have distinctive long white hairs</u> pointing backward toward the base of the stem (lower left). The alternate pale green 2-8" leaves feel rough when pulled through the fingers because of short, stiff hairs on their edges. Leaves at the tip of the stem point forward, often in a V or Y shape (middle). Flower clusters are <u>widely spreading panicles with sparse branches</u> (upper) bearing loosely <u>overlapping boat-shaped spikelets</u> lined up along them (lower right).

Identification: No other grass at FF/NW has the backward-pointing hairs on the nodes except for rice cutgrass *(Leersia oryzoides)*, uncommon at edges of the stream and marsh at FF. It is more robust with rougher leaves, larger spikelets and more branched panicles.

TIMOTHY
Phleum pratense
GRASS FAMILY *(Poaceae)*

Flowering Period: Jun - Jul.

Occurrence: <u>Roadsides, fields, woodland edges</u>, pastures, ditches. **FF:** Uncommon along left side of GM boardwalk just before reaching the blind.

Description: Introduced, bunched or single stemmed, erect perennial from 2-4' tall. <u>Stems have</u> a characteristic swollen or <u>bulblike base</u>. Leaves up to 12" long and 3/8" wide taper to a fine pointed tip. The <u>flower cluster</u> is an <u>erect cylindrical spike-like panicle</u> from 1-6" long. Tiny, short awns on the glumes of the crowded, single-flowered spikelets are responsible for its bristly surface.

Identification: Yellow foxtail *(Setaria glauca)* has an erect cylindrical flower cluster, but the bristles are longer and orange colored.

Comments: This species was a very early introduction from Europe. About 1720, Timothy Hanson of Maryland began to promote its use for hay, hence the common name "Timothy". Subsequently, it has been widely planted for hay and forage, especially in cool and moist regions. It is now naturalized throughout much of the U.S. and Canada.

JOHNSON GRASS
Sorghum halepense
GRASS FAMILY *(Poaceae)*

Flowering Period: Jun - Oct.

Occurrence: <u>Moist soils</u> in waste areas, ditches and field margins. **FF:** Rare; one of only two known colonies is on Hidden Lake Trail.

Description: Strongly rhizomatous introduced perennial with <u>stout stem up to 7'</u> <u>tall</u> growing in dense clumps or colonies. Large, wide, drooping <u>leaves</u> up to 36" long and 1/2 - 1 1/2" wide have a <u>prominent</u> <u>whitish midrib</u>. Flower cluster is a <u>large,</u> <u>open panicle</u> from 6-20" long usually with <u>whorled branches</u> and spikelets that are often tinged with red or purple.

Comments: This plant was introduced from the Mediterranean region around 1830 for use as a forage plant. It has moderate forage value, but contains hydrocyanic acid and may be poisonous to livestock if consumed when growth is interrupted by frost or drought. It is a serious weed in many places to our south, but it is not very tolerant of cold which inhibits its northward extension.

REDTOP
Agrostis gigantea; also
Agrostis stolonifera var. *major*
GRASS FAMILY *(Poaceae)*

Flowering Period: Late Jun - Jul.

Occurrence: <u>Low, moist bottomlands</u>. **FF:** Uncommon on North Stream Trail near junction with Cottonwood. **NW:** Rare on River Trail.

Description: Introduced rhizomatous perennial with flowering stems 1-3' tall. Stem, sheath and the narrow, pointed leaves are smooth and hairless. <u>Oval to pyramid-shaped, open panicles</u> up to 12" long have <u>whorled branches</u>, especially at the base. They bear many <u>red-to-purple spikelets</u>, each consisting of a single floret giving the panicle the color responsible for its common name. Color fades to gray after flowering.

Identification: The panicle shape is similar to Kentucky *(Poa pratensis)* and woodland bluegrass *(Poa sylvestris),* which flower earlier and do not have reddish spikelets.

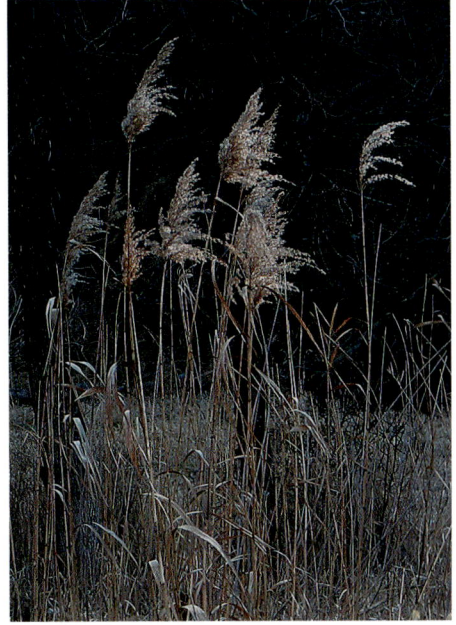

COMMON REED
Phragmites australis
GRASS FAMILY *(Poaceae)*

Flowering Period: Jun - Oct; mostly Jul - Aug.

Occurrence: <u>Wet soil in marshes</u>, ditches, margins of streams, lakes and ponds. **FF:** Common along edge of Great Marsh. **NW:** Uncommon in floodplain sites off-trail.

Description: <u>Our largest grass</u>, this perennial with stout, round stems up to 12' tall and 1" in diameter, has extensive, stout rhizomes which form <u>large colonies in wet areas</u>. The numerous smooth, flat leaf blades are up to 24" long and 1 1/2" wide. Flower clusters are <u>large (up to 14" long),</u> <u>dense</u>, light brown to purplish <u>panicles with ascending branches.</u> <u>As the plant matures the tips and branches nod.</u> Panicles develop a <u>plume-like, feathery appearance late in the season</u> when the long silky hairs on the tiny stems (rachillae) bearing the flowers become more visible (above right).

Comments: Common reed has a worldwide distribution. Populations in this country include both native and introduced plants. The aggressive behavior of some introduced strains has prompted its classification as a noxious weed in several states. Native Americans used common reed for thatching and mats and made arrow shafts from its stems. Other cultures have utilized the seeds, leaves and rhizomes for food. It provides good cover for birds and small mammals as well as nesting sites for waterfowl and small birds. The rhizomes are a food source for muskrats.

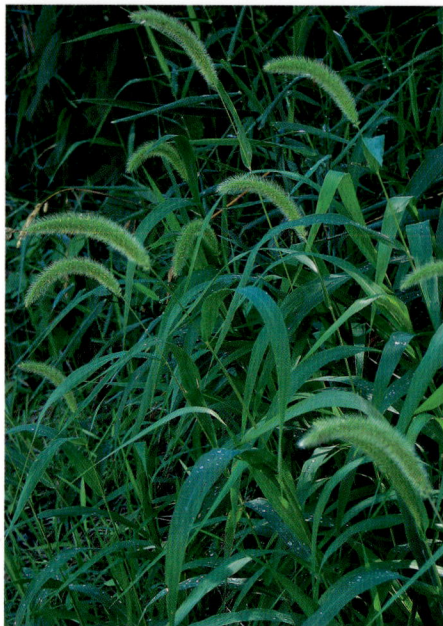

GREEN FOXTAIL
Setaria viridis
GRASS FAMILY *(Poaceae)*

Other Common Names: Bottlegrass, pigeongrass, green millet, wild millet, green bristlegrass.

Flowering Period: Jul - Sep.

Occurrence: Roadsides, disturbed areas. **FF:** Common along edges of Camp Gifford Road. **NW:** Common along edges of both parking lots.

Description: Weedy, introduced annual from 12-40" tall with erect or spreading stems which are often sharply angled at the base. Hairless leaf blades may be smooth or rough to the touch measuring up to 10" long and 1/2" wide. Hairs are present on the margins of the upper leaf sheath. The cylindrical 1-6" nodding (occasionally erect) flower cluster consists of densely packed spikelets, each associated with 1-3 pale green or purplish bristles (lower).

Identification: The flower spike of yellow foxtail *(S. glauca)* is erect and more cylindrical than other foxtails, and bristles are yellow or orange. Giant foxtail *(S. faberi)* has soft white hairs on the upper surface of the leaf (best seen with a hand lens). The upper leaf also often feels soft or velvety, particularly on those plants with many hairs. Less reliable, but often helpful, is the tendency for giant foxtail flower clusters to droop from the base making a broad, deep arch while green foxtail often just droops at the tip or is gently arched.

GIANT FOXTAIL
Setaria faberi
GRASS FAMILY *(Poaceae)*

Other Common Names: Chinese foxtail, Chinese millet, nodding foxtail, giant bristlegrass, pigeongrass.

Flowering Period: Jul - Oct.

Occurrence: Roadsides, disturbed sites, waste ground, cultivated fields, gardens. **FF:** Uncommon along margins of Camp Gifford Road parking lot. **NW:** Uncommon on edges of Jonas Center parking lot and in disturbed prairie sites.

Description: Stout, introduced, weedy, bunched annual from 1 1/2 - 4 1/2' tall with erect or leaning stems. Leaves are broad (up to 1/2") and may feel velvety because of the soft, white hairs that usually cover the upper surface. They are best seen when viewed tangentially (lower photo). The stem is smooth, but leaf sheath margins have fine hairs. The cylindrical 1-8" flower clusters usually nod from the base. They consist of many densely packed spikelets each equipped with several bristles responsible for its "foxtail-like" look.

Identification: See discussion under green foxtail *(S. viridis)* for separation from the other foxtails. Canada wild rye *(Elymus canadensis)* and hairy wild rye *(Elymus villosus)* have bristly, nodding seed heads, but the awns are longer and Canada wild rye awns are bent and twisted. Both have elongated "oat-like" seeds unlike the rounded foxtail seed.

167

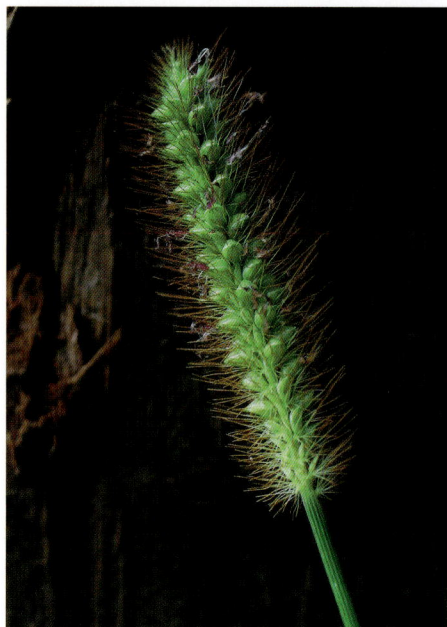

YELLOW FOXTAIL
Setaria pumila; also *Setaria glauca*
GRASS FAMILY *(Poaceae)*

Other Common Names: Yellow bristlegrass, wild millet, pigeongrass.

Flowering Period: Jul - Sep.

Occurrence: Roadsides, disturbed ground, fields and gardens. **FF:** Common along Camp Gifford Road. **NW:** Common at edge of Krimlofski parking lot and entrance trail.

Description: This foxtail is a weedy, introduced, bunched annual with erect-to-leaning stems 16-40" tall that are often sharply angled and reddish near the base. Leaf blades are loosely twisted or spiraled (above left). They may be hairless but often have long, thin hairs especially on the upper surface near the base (lower). The erect, cylindrical 1- 5" flower cluster bears many densely packed spikelets, each associated with 5-20 yellow bristles.

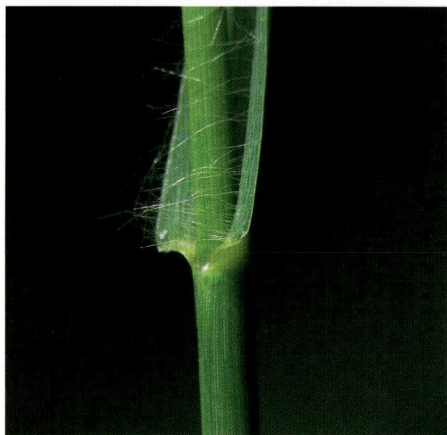

Identification: The erect flower head and yellow bristles separate this species from the other foxtails.

Comments: Natives of Eurasia, our foxtail species are among the most common and serious weeds found in cultivated fields. Seeds are a significant food source for songbirds and upland game birds.

BEARDED SPRANGLETOP
Leptochloa fusca ssp. *fascicularis*
GRASS FAMILY *(Poaceae)*

Flowering Period: Jul - Oct. Mostly Jul - Aug.

Occurrence: <u>Weedy plant in parking lots, roadsides</u>, gardens, moist soil in marshes, edges of streams, ponds and ditches. **FF:** Locally common on Camp Gifford Road near railroad crossing.

Description: Native weedy, bunched annual from 8-48" tall with stems that may be erect, leaning or almost flat on the ground. The <u>leaves</u> are flat, long (2-20") and narrow (up to 1/4") with a <u>prominent whitish midrib</u> (upper), especially near the base. Flower cluster is an <u>elongated</u> 4-20" <u>panicle</u> often

partly enclosed in the upper leaf sheath. Erect, <u>narrow, spike-like branches</u> bear numerous overlapping <u>spikelets</u> which hug the stem and are <u>lined up along its long axis</u> (lower).

SIDEOATS GRAMA
Bouteloua curtipendula
GRASS FAMILY *(Poaceae)*

Flowering Period: Jul - Sep.

Occurrence: Upland <u>prairies</u>, savannas and woodland openings. **FF:** Common in floodplain prairie planting. **NW:** Common in all prairie restorations; dominant in upper Jonas Prairie.

Description: Rhizomatous native perennial from 1-3' tall. Narrow, pointed leaves are up to 12" long and less than 1/4" wide, usually with sparse, long hairs on the edges near the base. The common name originates from the <u>erect or arching flower stems bearing short spikes aligned in 2 rows that hang downward along one side of the upper stem</u> (middle right). In flower, this otherwise inconspicuous grass has attractive showy reddish-orange anthers (lower).

Comments: Sideoats grama is a significant prairie species, especially in the drier uplands. An important range grass, it is a good source of forage for livestock and wildlife

BLUE GRAMA
Bouteloua gracilis
GRASS FAMILY *(Poaceae)*

Flowering Period: Jul - Aug.

Occurrence: Drier sites in upland prairies.
NW: Rare in Knull Prairie.

Description: Low growing, bunched or mat-forming native perennial. The 8-30" stems bear 1-3 comb-like flower clusters up to 1" long, the upper cluster always at the tip. Clusters consist of closely-packed flowers all arising from just one side of the secondary stem which is initially straight but curves upward slightly at maturity. Flowers extend out all the way to the tip of the stem.

Identification: The similar male flower cluster of buffalo grass *(Buchloe dactyloides)* is smaller and shorter stemmed. Buffalo grass also commonly produces above-ground runners called stolons. The stem holding the one-sided flowers projects well beyond the flower cluster in hairy grama *(B. hirsuta)*.

Comments: Blue grama is a common, often dominant, component of short grass prairies to the west.

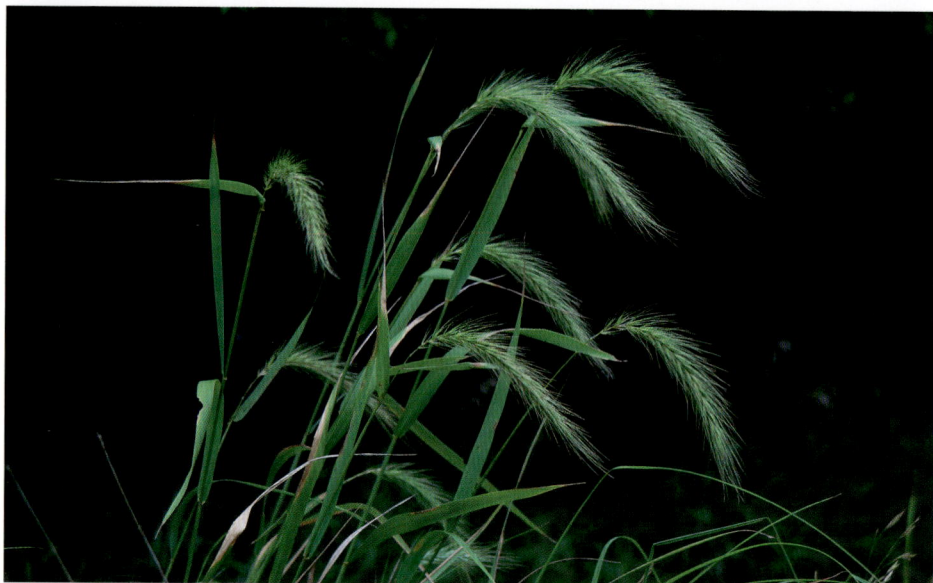

CANADA WILD RYE
Elymus canadensis
GRASS FAMILY *(Poaceae)*

Flowering Period: Jul - Aug.

Occurrence: <u>Prairies</u>, ditches, open woods and disturbed sites. **FF:** Uncommon in old prairie planting on History Trail. **NW:** Common in Knull and Nebraska Prairies.

Description: Native perennial growing in bunches from 3-5' tall. The stout stems, leaves and leaf sheaths are usually smooth and hairless. Broad <u>leaves</u> up to 15" long and 1/2" wide <u>have prominent auricles,</u> which clasp the back of the stem (lower). <u>Large arching flower spikes</u> are 3-7" long. The flowers have long awns which characteristically diverge or curve backward when they mature, giving the <u>spike</u> a <u>shaggy</u> appearance (middle).

Identification: See discussion under hairy wild rye *(E. villosus).*

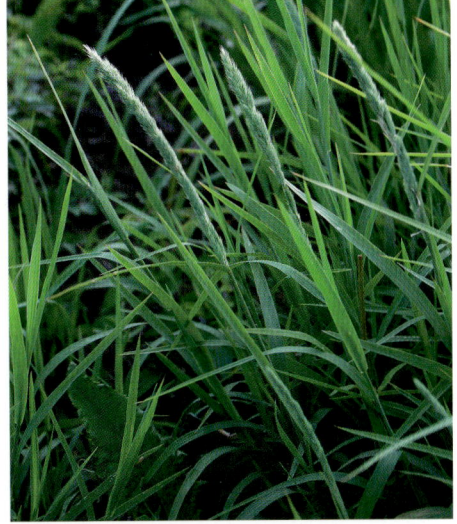

MARSH MUHLY
Muhlenbergia racemosa
GRASS FAMILY *(Poaceae)*

Other Common Name: Green muhly, wild timothy.

Flowering Period: Jul - Sep; mostly Jul-Aug.

Occurrence: Moist or dry open disturbed ground; roadsides; pastures. **FF:** Uncommon on Hawthorn Trail. **NW:** Uncommon in disturbed areas of Jonas and Nebraska Prairies.

Description: Native, erect rhizomatous perennial from 12-32" tall. Numerous smooth leaves arising all along the stem are up to 7" long and 1/4" wide. Panicles are closed, densely flowered, relatively thick, bristly, lobulated and from 1-5" long.

Identification: Other *Muhlenbergia* species are quite similar, usually (not always) separated by their more delicate, slender and less bristly flower heads. They more often occupy shadier habitats in woodlands or woodland openings.

STINKGRASS
Eragrostis cilianensis
GRASS FAMILY *(Poaceae)*

Flowering Period: Jul - Oct.

Occurrence: <u>Waste areas</u>, roadsides, lawns, fields. **FF:** Rare along Hidden Lake Trail. **NW:** Uncommon along edges of both parking lots.

Description: Bunched, weedy annual from 4-20" tall introduced from Europe. The erect, or more often leaning, flower stems are sharply bent and branched at the base. Narrow leaf blades are 2-8" long and less than 1/4" wide, often with a tuft of long hairs at the base where they join the sheath. The flower cluster is a dense 1-5" <u>distinctive grayish-green panicle bearing many conspicuous spikelets</u>. Each spikelet consists of 9-24 florets borne on a short pedicle (lower right).

Comments: Tiny resinous glands on the stems, leaf margins, glumes and lemmas are responsible for the disagreeable odor given off by the green plant when crushed.

CAROLINA LOVEGRASS
Eragrostis pectinacea
GRASS FAMILY *(Poaceae)*

Flowering Period: Jul - Oct.

Occurrence: <u>Waste areas</u>, roadsides, and fields. **NW:** Uncommon along edge of parking lot at CJIC.

Description: Weedy introduced annual from 4-24" tall. Flowering stems are erect or leaning, and lower stems are often sharply bent. Plants branch freely from the base, often forming dense bunches. The narrow leaves are 1-6" long and up to 1/4" wide. They are smooth except for small tufts of white hairs situated at the junction of the leaf base and sheath (collar). The flower cluster is a <u>delicate open panicle</u> from 2-6" long. Main branches are spreading. Secondary branches bearing the 6-13 flowered <u>spikelets</u> <u>are pressed flat against the main branches</u>, making the spikelets rather inconspicuous.

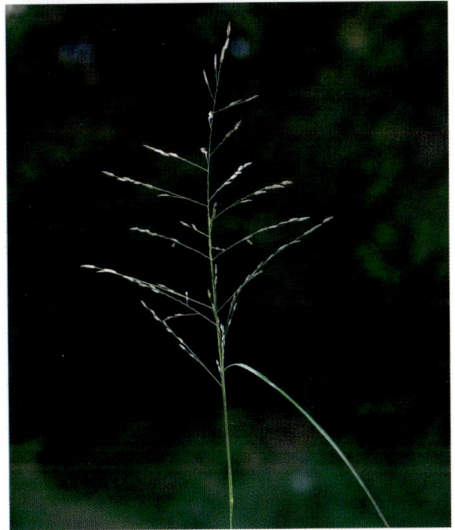

Identification: Stinkgrass *(E. cilianensis)* has a similar open panicle, but the larger, more conspicuous, grayish spikelets are spreading and not pressed flat against the main panicle branches

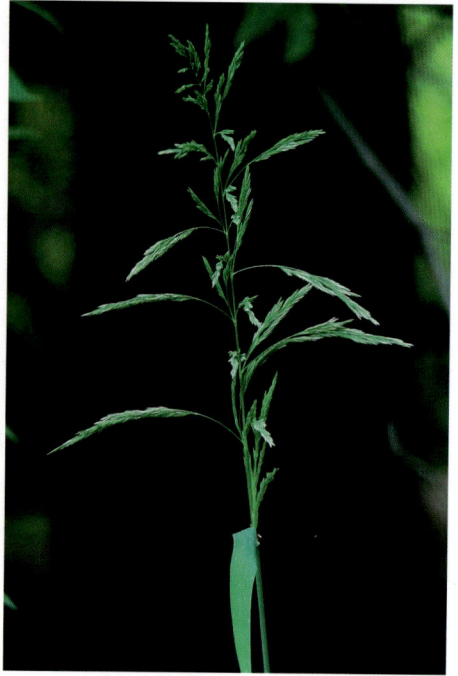

COMMON WOODREED
Cinna arundinacea
GRASS FAMILY *(Poaceae)*

Flowering Period: Jul - Sep.

Occurrence: <u>Floodplain woods</u>, <u>moist ravines</u>, ditches, streambanks. **FF:** Common on Cottonwood Trail. **NW:** Common on MRE Trail near Rock Creek.

Description: Native perennial occurring as a single plant or in small bunches. The <u>erect stems</u> are from 2-5' tall. Leaves up to 14" long and 1/2" wide with roughened margins are rather loosely spaced along the entire stem. <u>Leaf sheaths and stems are smooth and hairless</u>. The flower cluster is a <u>dense, erect panicle</u> from 4-10" long. Except for a short section near the base, the <u>ascending branches</u> are covered with closely packed spikelets. The base of the panicle is often enclosed within the upper leaf sheath.

Identification: Ear-leaf brome *(Bromus latiglumis)*, found in similar moist habitats, also flowers at the same time. It has a drooping panicle, leaf sheaths covered with fine hairs, and conspicuous projections (auricles) where the leaf base joins the stem.

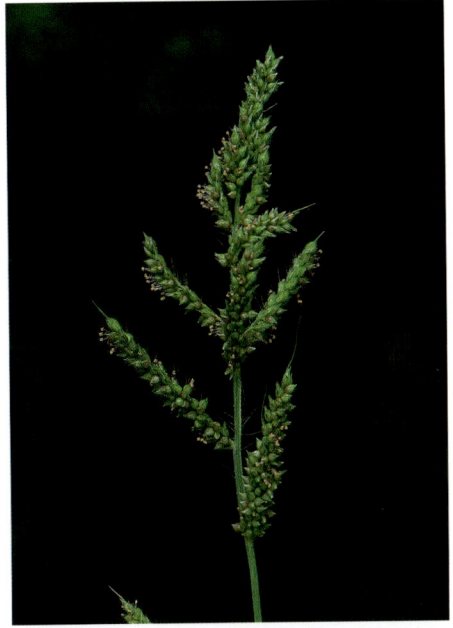

BARNYARD GRASS
Echinochloa crusgalli
GRASS FAMILY *(Poaceae)*

Other Common Names: Wild millet, cockspur, watergrass.

Flowering Period: Jul - Sep.

Occurrence: Moist disturbed areas. **FF:** Locally common on edge of Great Marsh east of the blind. **NW:** Locally common on edges of pond in Jonas Valley.

Description: An annual introduced from Eurasia, this weedy plant has a stout reclining-to-erect stem from 1 to 3 1/2' tall. The flat, usually hairless spreading leaves are from 3-12" long and up to 1/2" wide with a conspicuous white midvein. The flower cluster is an erect or nodding green-to-purple panicle from 2-10" long with 5-12 appressed-to-spreading branches. The densely flowered branches bear large numbers of overlapping spikelets arranged

along one side of the stem and awns or bristles of variable length and density (lower).

Identification: A species which is native to North America, rough barnyard grass *(E. muricata),* occupies similar habitats and is separated only by microscopic differences in the tip of the fertile lemma. It grows at NW on MRE Trail where the bridge crosses Rock Creek.

Comments: Barnyard grass is a prolific seed producer (up to 40,000 per plant) and an important food source for birds, especially waterfowl. The seeds of the native species were a food source for Native Americans.

BIG BLUESTEM
Andropogon gerardii ssp. *gerardii*
GRASS FAMILY *(Poaceae)*

Other Common Name: Turkeyfoot.

Flowering Period: Jul - Oct.

Occurrence: Upland and lowland prairies; often dominant in lowland sites; roadsides.
FF: Common in floodplain prairie planting.
NW: Common in all prairie restorations.

Description: Native 3-6' perennial, occasionally as tall as 9', growing in bunches or spreading by rhizomes to form turfs. The numerous leaves are up to 20" long and less than 1/2" wide, usually with a few long hairs near the base. Stout, rounded stems often have areas of bluish or purple coloration (lower right). At the tip of the stem are clusters of 2-7 linear flower stalks from 2-4" long. Each stalk originates from a common point, the cluster resembling a "turkeyfoot" (middle right). Stems turn a rich reddish-brown at maturity.

Comments: Big bluestem was the dominant grass in much of the tallgrass prairie which once covered most of the FF/NW vicinity. The Omaha-Ponca laid stems of "red hay" over the wooden framework of their lodges for support before covering them with earth.

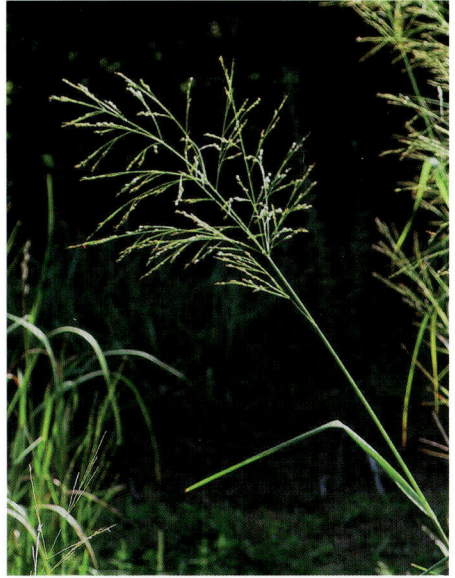

SWITCHGRASS
Panicum virgatum
GRASS FAMILY *(Poaceae)*

Flowering Period: Late Jul - Sep.

Occurrence: <u>Prairie</u> and open woodlands. Prefers moist soil, but found in drier sites as well. **FF:** Locally common in old prairie planting on History Trail. **NW:** Common in Knull, Koley and Lower Jonas Prairies.

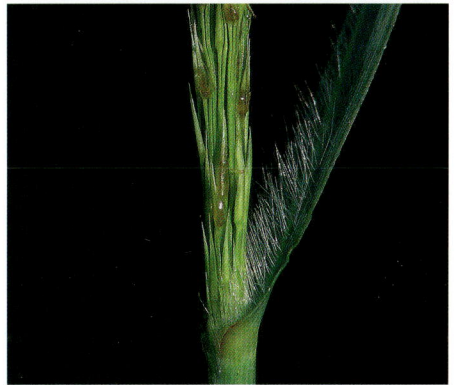

Description: Strongly rhizomatous, sod-forming native perennial often occurring in large clumps. Flowering stems are 3-6' tall. Abundant leaves up to 24" long and 3/4" wide have a <u>characteristic zone of long, dense hairs on the upper leaf</u> just above the junction with the sheath (middle right). The flower cluster is a <u>large 6-18" long panicle with many open, spreading branches</u>. At their tips are groups of <u>simple</u>, teardrop-shaped, green, red or purple <u>spikelets</u> which fade to a light brown after flowering.

Comments: Switchgrass was an important component of our tallgrass prairies. It is a potential biomass energy crop, and cultivars are popular garden ornamentals.

PRAIRIE CORDGRASS
Spartina pectinata
GRASS FAMILY *(Poaceae)*

Other Common Names: Sloughgrass, ripgut, tall marshgrass.

Flowering Period: Late Jul - Sep.

Occurrence: Wet prairies, marshy areas, moist sites along ponds, streams and ditches. **FF:** Locally common in floodplain prairie planting. **NW:** Locally common in prairie transplant in front of CJIC.

Description: Native perennial from 3-8' tall with extensive stout rhizomes, often forming dense stands in wet areas. Firm leaf blades up to 48" long and 1/2" wide have abrasive roughened or toothed margins. The flower cluster is a tight, dense panicle with 5-30 ascending branches bearing 2 rows of flower spikelets in a comb-like arrangement along one side of the branch.

Identification: Other prairie species including buffalo grass *(Buchloe dactyloides)* and grama grasses in the genus *Bouteloua* have comb-like, one-sided flower clusters, but individual clusters are widely separated and not configured in a dense, branching panicle like cordgrass. They, also, are shorter (less than 3' tall) and occupy drier upland sites.

Comments: Cordgrass, a dominant plant in moister portions of the tallgrass prairie, was used by Native Americans who thatched the wooden framework of their lodges before covering them with earth. The sharp leaf margins can cut exposed skin, hence the common name "ripgut".

TALL WHEATGRASS
Elymus elongatus; also *Thinopyrum ponticum; Agropyron elongatum*
GRASS FAMILY *(Poaceae)*

Flowering Period: Jul - Sep.

Occurrence: <u>Moist</u>, especially alkaline <u>soils</u>, waste ground, roadsides. **FF:** Uncommon and planted on Childs Hollow Dam where it was in the grass seed mix used to cover the dam surface.

Description: Introduced, nonrhizomatous perennial with flowering stems from 3 1/2 to 6 1/2' tall. Narrow leaves up to 24" long have rolled up edges and 6-8 prominent ridges on the upper side. A white, waxy substance on the surface often gives the leaves a distinct bluish coloration. Auricles are prominent. Flower cluster is an <u>open spike</u> up to 18" long with <u>little or no overlapping of individual spikelets</u> (lower right). Each spikelet has 5-11 florets with round to pointed tips and blunt-tipped glumes at its base (lower left).

Identification: Spikes of quackgrass *(Elymus repens)* and western wheatgrass

(Elymus smithii) have densely overlapping spikelets, and their glumes have pointed tips. Quackgrass leaves are not prominently ridged.

Comments: Tall wheatgrass is an important source of genes for wheat breeders, especially for stem and leaf rust resistance.

WILD RICE
Zizania palustrus
GRASS FAMILY *(Poaceae)*

Other Common Names: Indian wild rice, blackbird oats, marsh oats.

Flowering Period: Jul - Aug.

Occurrence: Ditches and <u>edges of ponds, streams and marshes, often in shallow water</u>. **FF:** Rare in Handsome Hollow near RR crossing. Neither author has seen this plant since the above photo was taken in 2002.

Description: Native, stout-stemmed, solitary annuals <u>up to 9' tall</u>. Strap-like leaves are up to 36" long and 1 1/2" wide. <u>Unisexual</u> <u>flowers</u> occur on open, many-branched pyramidal panicles up to 24" long and 12" wide. <u>Male flowers are on lower branches which spread widely</u> and droop at maturity. The numerous single-flowered <u>female spikelets are on erect branches at the top</u>.

Comments: Wild rice was an important food source for Native Americans of the Great Lakes region who boiled and ate it with beans, corn or squash. Once considered a gourmet food, it is now cultivated and available in many food stores. Wild rice is an important wildlife food source for waterfowl, particularly Mallards, Black and Wood Ducks as well as songbirds including Bobolinks and Red-winged Blackbirds.

GOOSEGRASS
Eleusine indica
GRASS FAMILY *(Poaceae)*

Other Common Names: Wiregrass, silver crabgrass, crowsfoot grass.

Flowering Period: Late Jul - Oct.

Occurrence: <u>Compacted soil along roadsides and trails</u>, waste areas, pastures and lawns.
FF: Common in Gifford Road parking lot.
NW: Uncommon along edges of parking lots.

Description: Weedy annual introduced from warmer portions of Eurasia. The reclining, leaning or erect stems are 4-30" long and often white or silvery at the base. Leaves are up to 1/4" wide and 2-6" long. At the tip of the stem are <u>1-10 spreading, spikelike, whorled branches</u> (right middle) often with one or two additional branches coming off below the whorl. They are from 1 1/2 - 3" long with <u>flattened stems bearing 2 crowded rows of spikelets along one side</u> (lower photo, right side).

Identification: Crabgrasses *(Digitaria species)* often root at the nodes and don't

grow as a single, firmly rooted bunch like goosegrass. The branches of crabgrass (lower photo, left side) are thinner and the spikelets with their single fertile floret much smaller than the goosegrass spikelets (lower photo, right side), which have 2-9 florets.

183

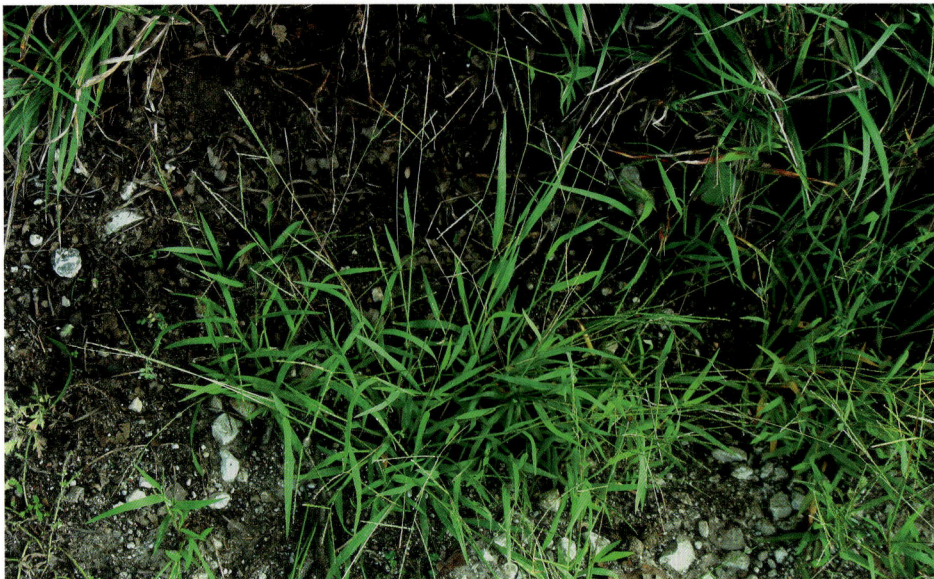

SMOOTH CRABGRASS
Digitaria ischaemum
GRASS FAMILY *(Poaceae)*

Flowering Time: Aug - Oct.

Occurrence: <u>Roadsides, trails</u>, disturbed areas, lawns, gardens, fields. **FF:** Common along Marsh and South Stream Trails. **NW:** Common on and along the edges of MRE Trail.

Description: This species, an annual introduced from Eurasia, is by far the most common crabgrass at FF/NW. It may be upright or spreading and often roots at the lower nodes. The flat, 1-5" long leaf blades have no midrib and are hairless or have just a few long hairs near the base. <u>Leaf sheaths</u> are flattened (keel shaped) and <u>hairless</u> (lower right). Flower stems have <u>2-8 flattened branches all arising from one point</u> (digitate) or along the stem within a short distance of each other. Branches bear two rows of spikelets on short stalks.

Identification: See discussion in goosegrass *(Eleusine indica)* and southern crabgrass *(Digitaria ciliaris)*.

Comments: These tough plants are a serious weed in many area lawns.

SOUTHERN CRABGRASS
Digitaria ciliaris
GRASS FAMILY *(Poaceae)*

Flowering Period: Aug - Oct.

Occurrence: <u>Trail edges</u>, lawns, gardens, fields, roadsides, waste areas. **FF:** Uncommon on Hackberry Trail below the outdoor patio. **NW:** Locally common on Jonas Trail in Raccoon Hollow.

Description: Introduced annual forming loose bunches or mats. Flowering stems up to 2 1/2' tall are erect, ascending or reclining and often root at the nodes. Flat leaf blades have prominent midribs and are usually hairless or have just a few hairs near the base. <u>Leaf sheaths</u> are keel shaped and <u>covered with long, soft white hairs</u> (lower right). Flower stems have 3-10 flattened <u>branches arising from a single point</u> <u>(digitate) or along the stem within a short</u> <u>distance of each other</u>. Each branch bears 2 rows of spikelets on short stalks.

Identification: Smooth crabgrass *(Digitaria ischaemum)* has hairless leaf sheaths. It is by far the most common crabgrass species along our trails. Another very similar species, hairy crabgrass *(Digitaria sanguinalis),* was not identified by the authors but is to be expected at FF/NW. It has more hair on the leaves, but other characteristics overlap those of southern crabgrass, and these two species cannot be reliably separated without a microscope.

WIRESTEM MUHLY
Muhlenbergia frondosa
GRASS FAMILY *(Poaceae)*

Flowering Period: Late Jul - Oct.

Occurrence: <u>Upland and floodplain woods and edges</u>. **FF:** Common on Hackberry, Hawthorn and Hidden Lake Trails. **NW:** Common at Krimlofski entrance and on MRE Trail.

Description: Strongly rhizomatous perennial often forming <u>large colonies</u>. The rather <u>weak</u> ascending or erect 1-3' flowering <u>stems</u> usually lean noticeably or lie almost flat at maturity. <u>Compact to slightly open 1-5" panicles</u> arise from the tip of the stem and from side branches which are especially numerous in this species. Panicles of side branches are often partially enclosed in the leaf sheaths. Stems and nodes are hairless, smooth and shiny. Plants have <u>abundant hairless leaves</u> usually 1-6" long and less than 1/4" wide.

Identification: Whitegrass *(Leersia virginica)* also flowers late, has weak stems and forms large clumps or colonies. It has a much different open, sparsely branched panicle and prominent white hairs on the nodes.

Comments: Three other species of *Muhlenbergia* are quite common at FF/NW. These "look-alikes" possess either obscure or overlapping characteristics that make separation unreliable without a microscope. They are Mexican muhly *(M. mexicana),* shown in the middle and lower right photos on the opposite page; Bush's or nodding muhly *(M. bushii);* and forest muhly *(M. sylvatica),* whose flower cluster is shown on the lower left, opposite. As noted, these are distinct species, but require separation using characteristics of the ligules, internodes and glumes, which are beyond the scope of this book.

NIMBLEWILL
Muhlenbergia schreberi
GRASS FAMILY *(Poaceae)*

Flowering Period: Aug - Oct.

Occurrence: <u>Shady disturbed areas,</u> <u>trail edges</u>, moist woods, roadsides, fields, lawns, gardens. **FF:** Common on trail edges in Childs Hollow. **NW:** Common along edges of upper Raccoon Hollow Trail (lower photo).

Description: Native, non-rhizomatous perennial from 8-30", <u>usually less than 12" tall</u>, growing in bunches or extensive colonies. Stems are <u>erect early</u> in the season, but <u>by fall many sprawl</u> along the ground, often forming new roots at the nodes. Leaves are short (less than 4" long), narrow and smooth except for hairs near the junction of the leaf and stem. The <u>flower cluster is a very slender, condensed panicle</u> 2 1/2 - 9" long, its base sometimes partially included in the leaf sheath.

Comments: Nimblewill can be a very aggravating and hard-to-eradicate weed in lawns and gardens.

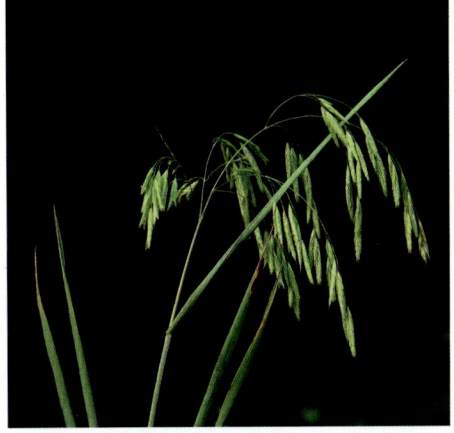

EAR-LEAF BROME
Bromus latiglumis
GRASS FAMILY *(Poaceae)*

Flowering Period: Aug - Oct.

Occurrence: <u>Moist woodlands</u>, both upland and floodplain, but more common on floodplain. **FF:** Common on Missouri Trail near river. **NW:** Locally common in opening near the south side of the pond on River Road.

Description: Native perennial from 2 1/2 - 5' tall growing singly or in bunches, often in large, loosely-spaced groups. <u>Leaves are numerous</u>, usually more than 10. Their overlapping <u>sheaths, usually covered with dense, fine grayish hairs,</u> (middle) completely cover the underlying smooth stem and nodes. Projections called <u>auricles extend backward at the base of the leaf</u> (lower). The upright flowering stem has a moderately dense <u>drooping 6-10" long panicle</u> at its tip.

Identification: Common woodreed *(Cinna arundinacea)* occupies similar sites and flowers at the same time, but has an erect panicle, hairless leaf sheaths and no auricles. Canada brome *(Bromus pubescens)* flowers much earlier

(May/June), has fewer leaves (less than 10 per stem) and no auricles. Smooth brome *(Bromus inermis)* also flowers earlier, has an upright panicle and smooth, hairless stems and sheaths.

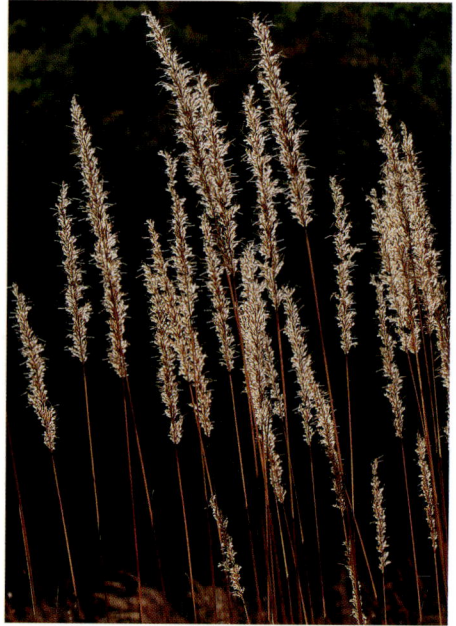

INDIAN GRASS
Sorghastrum nutans
GRASS FAMILY *(Poaceae)*

Flowering Period: Aug - Sep.

Occurrence: <u>Prairies</u>, open woods, fields. Prefers deep moist soils; also found in uplands. **FF:** Common in floodplain prairie planting. **NW:** Common in all prairie restorations.

Description: Native perennial from 2-7' tall with short rhizomes, often forming clumps. Pale green leaves are up to 18" long and less than 1/2" wide. At the base of the leaf where it joins the stem are a pair of narrow, pointed <u>"rabbit-ear" projections</u> (lower left). The flower cluster is a fairly dense 6-12" long <u>lance-shaped panicle</u>. Initially, it is golden-yellow but fades to grayish-brown. <u>Florets have bent and twisted bristle-like awns</u> (lower right) and are covered with whitish hairs that glisten in the autumn sun. Flowering plants have prominent golden yellow anthers.

Comments: Indian grass is one of the principal species of the tallgrass prairie often growing in association with big bluestem.

PURPLETOP
Tridens flavus
GRASS FAMILY *(Poaceae)*

Other Common Name: Redtop.

Flowering Period: Aug - Oct.

Occurrence: Upland prairies, open woods, pastures, roadsides. **FF:** Uncommon on South Stream Trail. **NW:** Uncommon in Nebraska Prairie.

Description: Native bunched perennial with stout flowering stems from 2-5' tall. Green stems have tinges of red about the dark reddish nodes (lower right). Leaf blades are 8-20" long and less than 1/2" wide. Upper leaf sheaths have a focal collection of hairs where they join the leaf base (lower left). The flower cluster is an open panicle with erect or drooping branches bearing reddish to purplish spikelets at their tips.

Identification: Indiangrass *(Sorghastrum nutans)* has long awns and brown spikelets. Switchgrass *(Panicum virgatum)* has a focal collection of hairs on the top side of the lower leaf and an open, erect panicle. Redtop *(Agrostis gigantea)* has red spikelets but is shorter, flowers earlier and has whorled branches.

WITCHGRASS
Panicum capillare
GRASS FAMILY *(Poaceae)*

Other Common Names: Tickle or tumbleweed grass, witch's hair.

Flowering Period: Aug - Oct.

Occurrence: Disturbed ground, fields, roadsides. **NW:** Locally common in disturbed site in Jonas Prairie (above right). Rare elsewhere.

Description: Native, bunched weedy annual up to 28" tall. Leaves are up to 10" long and 1/2" wide. Often reddish or purple, the leaf sheaths have numerous long hairs (lower right). Mature flower clusters are large, open, spreading, much branched panicles up to 12" long and almost as wide, which often are reddish or purplish (upper right). Panicles break away from the plant, rolling off as tumbleweeds at maturity.

Identification: Fall panicum *(P. dichotomiflorum)* and switchgrass *(P. virgatum)* have open panicles but do not have hairs on the leaf sheath

FALL PANICUM
Panicum dichotomiflorum
GRASS FAMILY *(Poaceae)*

Other Common Names: Zigzag grass, western witchgrass.

Flowering Period: Aug - Oct.

Occurrence: <u>Moist disturbed areas, roadsides</u>, fields. **FF:** Uncommon along floodplain portion of Camp Gifford Road. **NW:** Uncommon at edge of Krimlofski parking lot.

Description: Native, bunched, weedy annual with stout-looking but rather weak leaning stems from 1-3' tall. <u>Stems are often slightly bent at the swollen nodes giving them a zigzag appearance</u> (lower right). The stem, broad leaves (up to 3/4" wide) and <u>leaf sheath</u> are <u>smooth and hairless</u>. Flower clusters are broad <u>open panicles</u>. Axillary panicles occur at each upper node. The largest panicle is at the tip, measuring up to 14" long. Outer branches appear thickened

because the spikelets are lined up along and quite closely appressed to the secondary panicle branches.

Identification: Fall panicum shares a broad, open panicle with witchgrass *(P. capillare)* and switchgrass *(P. virgatum)*. Neither has swollen nodes or zigzag stems. Witchgrass also has prominent hairs on the stem and sheath while those of fall panicum are hairless.

LITTLE BLUESTEM
Schizachyrium scoparium; also
Andropogon scoparius
GRASS FAMILY *(Poaceae)*

Flowering Period: Aug - Oct.

Occurrence: <u>Upland prairies</u>, savannas. **FF:** Uncommon on dam in Childs Hollow. **NW:** Common in Jonas and Koley Prairies.

Description: <u>Mid-sized</u> native perennial bunchgrass with flowering stems from 1-4' tall. Narrow leaves are up to 12" long and less than 1/4" wide. <u>Leaves and stems are often bluish.</u> Stems are often flattened at the base, a feature accentuated by the <u>sharply keeled leaf sheaths</u>. Flowering stems have numerous branches, each with its own <u>unique 1-2" long flower cluster</u> at the tip consisting of several paired spikelets, one fertile and the other infertile. At maturity the long white hairs on the stalk of the infertile spikelet and pedicle give the plant a festive <u>feathery appearance</u> (lower right). <u>Foliage turns a rich red brown or bronze</u> in the fall.

Comments: Little bluestem, the official state grass of Nebraska, is an important component of our native tallgrass and mixed grass prairies thriving on drier sites in a wide variety of soil types. The bluish foliage and rich reddish-brown fall color have made it a popular ornamental.

GRASSES LISTED BY HABITAT

GRASSES OF THE UPLAND WOODS
Kentucky Bluegrass
Woodland Bluegrass
Nodding Fescue
Hairy Wild Rye
Bottlebrush Grass
Canada Brome
Bearded Wood Grass
Black-seed Ricegrass
American Beakgrain
Whitegrass
Wirestem Muhly (and other Muhly species)
Ear-leaf Brome

FLOODPLAIN GRASSES
Kentucky Bluegrass
Woodland Bluegrass
Slender Wedgegrass
Nodding Fescue
Fowl Mannagrass
Reed Canary Grass
Canada Bluegrass
Hairy Wild Rye
Virginia Wild Rye
Bottlebrush Grass
American Beakgrain
Whitegrass
Redtop
Canada Brome
Common Reed
Common Woodreed
Wild Rice
Wirestem Muhly (and other Muhly species)
Ear-leaf Brome

SPECIES FOUND FREQUENTLY ON TRAILS AND TRAIL EDGES
Annual Bluegrass
Kentucky Bluegrass
Whitegrass
Smooth Crabgrass
Southern Crabgrass
Nimblewill
Path Rush*

COMMENTS
*As the name suggests, this grass-like plant is a rush, not a grass, found in compacted soil like the grasses listed here.

Remember that on occasion a grass will be found oustide its "typical" habitat.

PRAIRIE GRASSES

NW Sites
Transplant in front of CJIC
Knull Prairie restoration
Koley Prairie restoration
Millard Prairie Transplant
Jonas Valley Restoration
Nebraska Prairie Restoration

FF Sites
Floodplain prairie planting east of blind
Planting on Childs Hollow Dam
Old planting on History Trail
Buffalo Grass
Scribner's Panicgrass
Western Wheatgrass
Eastern Gama Grass
Porcupine Grass
Sideoats Grama
Blue Grama
Canada Wild Rye
Big Bluestem
Switchgrass
Prairie Cordgrass
Indian Grass
Purpletop
Little Bluestem

'WEEDY' GRASSES
Plants of old fields, woodland edges, disturbed ground in prairies and woodlands, roadsides, ditches, parking lots
Orchard Grass
Reed Canary Grass
Smooth Brome
Japanese Brome
Tall Fescue
Little Barley
Foxtail Barley
Quackgrass
Timothy
Johnson Grass
Green Foxtail
Giant Foxtail
Yellow Foxtail
Bearded Sprangletop
Marsh Muhly
Stinkgrass
Carolina Lovegrass
Barnyard Grass
Goosegrass
Smooth Crabgrass
Southern Crabgrass
Witchgrass
Fall Panicum

GLOSSARY

Achene: A hard, one-seeded fruit that does not split open.

Alternate (leaves): Placed singly, not in pairs or whorls, on stems.

Annual: Completing its life cycle in one year.

Anther: The pollen-bearing, enlarged end of the stamen.

Appressed: Closely pressed against.

Auricle (grass): An ear-shaped appendage at the base of a leaf blade (see Illustrated Glossary).

Awn: A bristle-like appendage (see Illustrated Glossary).

Basal (leaves): Rising directly from the roots.

Beak (sedge): The narrowed tip of the perigynium (see Illustrated Glossary).

Biennial: Completing its life cycle in two years.

Bipinnate (leaves): Twice compound (see Illustrated Glossary).

Blade: The broad, expanded part of the leaf.

Bloom: A white powdery coating on shiny stems, fruits or leaves.

Bract: A modified leaf, usually at the base of a flower.

Calyx: The outer ring of sepals.

Capsule: A dry fruit, which splits to yield 2 or more seeds.

Catkin: A dense, often drooping flower cluster spike.

Compound (leaves): Divided into two or more parts (leaflets).

Cordate (leaf): Heart-shaped (see Illustrated Glossary).

Corolla: The inner envelope of flower petals.

Corymb: A flat cluster of flowers with those at the margin flowering first (see Illustrated Glossary).

Culm: A flower-bearing stem of a grass or sedge (see Illustrated Glossary).

Cyme: A cluster of flowers terminating in a central flower which blooms first (see Illustrated Glossary).

Dioecious: Having male and female flowers on different plants.

Dissected (leaves): Finely cut or divided into narrow segments (see Illustrated Glossary).

Drupe: A fleshy fruit with one seed.

Elliptic (leaf): Oval in shape; widest at its middle (see Illustrated Glossary).

Entire (leaf): Having a smooth edge, not toothed or lobed (see Illustrated Glossary).

Fascicle: A bundle.

Family (plant):	A group of closely related genera.
Filament:	The thread-like stem of a stamen.
Floret (grass):	A grass flower (see Illustrated Glossary).
Fruit:	The ripened ovary containing the seed(s).
Genus:	A group of closely related species (plural: genera).
Glume (grass):	One of a pair of bracts at the base of a grass spikelet (see Illustrated Glossary).
Herb:	A plant which dies back to the ground in winter.
Hybrid:	Resulting from interbreeding of species.
Inflorescence:	The flowering portion of a plant.
Introduced:	Intentionally or accidentally established by man.
Keel (grass):	A prominent ridge along the back of the leaf or sheath.
Key:	A one-seeded fruit with a wing; a samara.
Lanceolate (leaf):	Lance-shaped, much longer than wide (see Illustrated Glossary).
Leaflet:	One part of a compound leaf.
Legume:	A member of the bean family (see Illustrated Glossary).
Lemma (grass):	The lower of two bracts enclosing a grass flower (see Illustrated Glossary).
Ligule:	The membrane at the junction of the blade and sheath in grasses and sedges (see Illustrated Glossary).
Linear (leaf):	Long, narrow with sides nearly parallel (see llustrated Glossary).
Lobed (leaf):	Edges deeply divided (see Illustrated Glossary).
Monoecious:	Having separate male and female flowers on the same plant.
Native:	Indigenous to an area; occurring naturally, not introduced by man.
Naturalized:	Not native to an area; introduced and reproducing on its own.
Nerve:	A prominent vein or rib on a leaflet, glume or perigynium.
Node:	A joint on the stem where a leaf is attached.
Oblong (leaf):	Longer than broad, with sides nearly parallel (see Illustrated Glossary).
Ovate (leaf):	Shaped like a pointed egg (see Illustrated Glossary).
Palea (grass):	The upper of two bracts enclosing a grass flower (see Illustrated Glossary).
Palmate (leaflets):	Attached to a single point on a stalk (see Illustrated Glossary).
Panicle:	A branched flower cluster with individual flowers attached to its branches, not the main stem (see Illustrated Glossary).
Pedicel:	The stalk of a single flower.
Perennial:	Completing its life cycle in 3 or more years.
Perfoliate (leaf):	Completely surrounding the stem.
Perigynium (sedge):	A saclike structure enclosing the ovary in the genus Carex (see Illustrated Glossary).
Petal:	Individual part of the corolla often brightly colored; may be separate or fused to form funnels or bells.
Petiole:	A leaf stalk.

Pinnate (leaf):	Leaflets arranged in two rows on a stalk (see Illustrated Glossary).
Pistil:	The female organ of a plant, consisting of ovary, style and stigma (see Illustrated Glossary).
Prickle:	A small, pointed extension of the plant's skin.
Pubescent:	Covered with short, downy hairs.
Raceme:	A slender flower cluster; each flower has its own stalk (see Illustrated Glossary).
Rachis:	The axis of a compound inflorescence.
Rhizome:	An underground stem.
Recurved:	Curved backward or downward.
Root Sucker:	A shoot which grows from the roots of a plant.
Root Sprout:	A shoot which grows from the roots of a cut tree.
Rosette:	Arranged in a compact circle.
Sepal:	A leaf-like division of the calyx; usually green.
Sessile:	Not stalked.
Samara:	A winged fruit.
Scale (sedge):	A dry, thin bract associated with both male and female flowers (see Illustrated Glossary).
Sheath:	That portion of a leaf which surrounds the stem of a grass or sedge(see Illustrated Glossary).
Spatulate (leaf):	Spatula or spoon-shaped (see Illustrated Glossary).
Spike:	A slender stem with a cluster of flowers, lacking individual stalks.
Spikelet (grass):	The basic flowering unit in grasses, consisting of two glumes at the base plus one to many florets.
Spine:	A sharp-pointed modification of a branch.
Stamen:	The male organ, consisting of filament and pollen-producing anther (see Illustrated Glossary).
Stigma:	The tip of the pistil, which is usually enlarged; it receives the pollen (see Illustrated Glossary).
Stipule:	A small leaf-like appendage at the base of a leaf stalk.
Stolon (grass):	A creeping, above-ground stem which forms roots at its nodes.
Style:	The stalk connecting the ovary with the stigma.
Tendril:	A slender, twisting organ used by a plant to gain support.
Tepal:	Used to describe sepals and petals when they are similar.
Thorn:	A modified stem which terminates in a point.
Trifoliate:	Three leaflets attached to the same leaf stalk.
Thyrse:	A compact flower cluster; each flower has its own stalk (see Illustrated Glossary).
Umbel:	A flower cluster with flower stalks radiating from a single point (see Illustrated Glossary).
Whorled (leaves):	Three or more leaves from the same node on a stem (see Illustrated Glossary).
Wing:	Thin flaps extending from stems, leaf stalks or seeds.
Woody (vines):	Persisting through winter; i.e., not dying back like herbs.

Flower Types

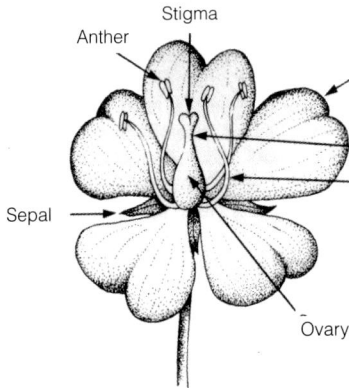

Simple flower

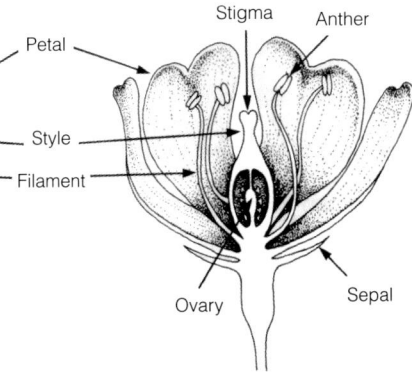

Simple flower (cross-section)

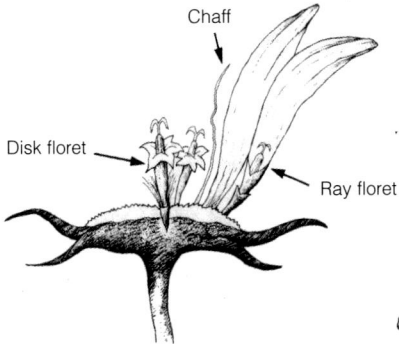

Composite flower (cross-section)

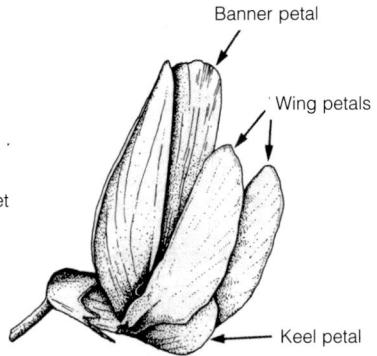

Legume flower

Inflorescence Types

Raceme

Spike

Thyrse

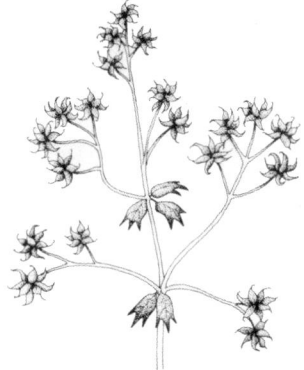

Panicle

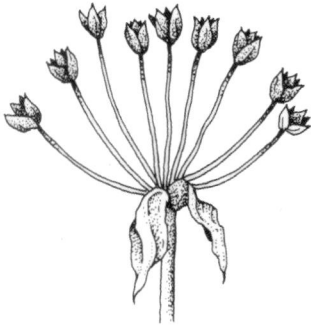

Umbel

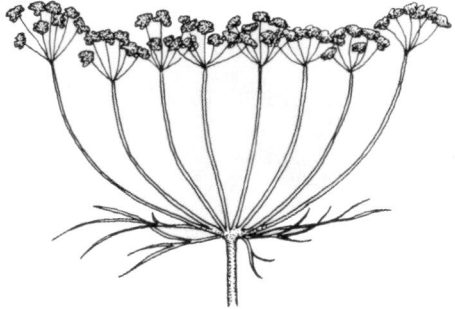

Compound Umbel

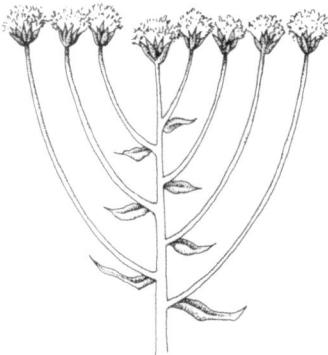

Corymb

Cyme

Leaf Shapes

Linear   Lanceolate   Spatulate   Oblong   Elliptic   Ovate   Cordate

Leaf Margins

Entire   Toothed   Lobed   Dissected

Leaf Types

Simple   Palmate   Pinnately Compound   Bipinnately Compound

Leaf Arrangements

Opposite   Alternate   Whorled

203

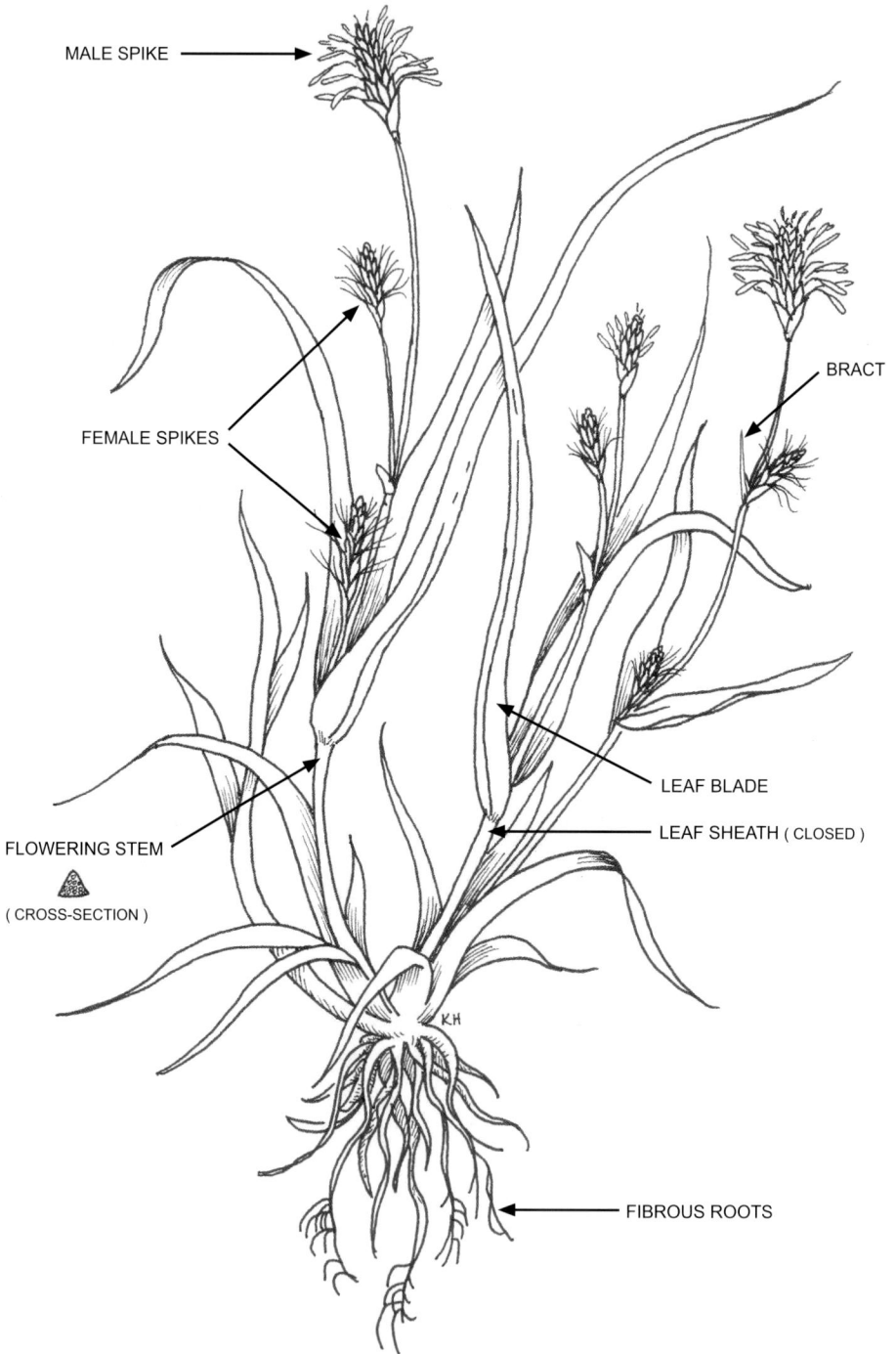

MALE SPIKE

FEMALE SPIKES

BRACT

FLOWERING STEM

(CROSS-SECTION)

LEAF BLADE

LEAF SHEATH (CLOSED)

FIBROUS ROOTS

K.H

SEDGE
(CAREX SPECIES WITH UNISEXUAL SPIKES)

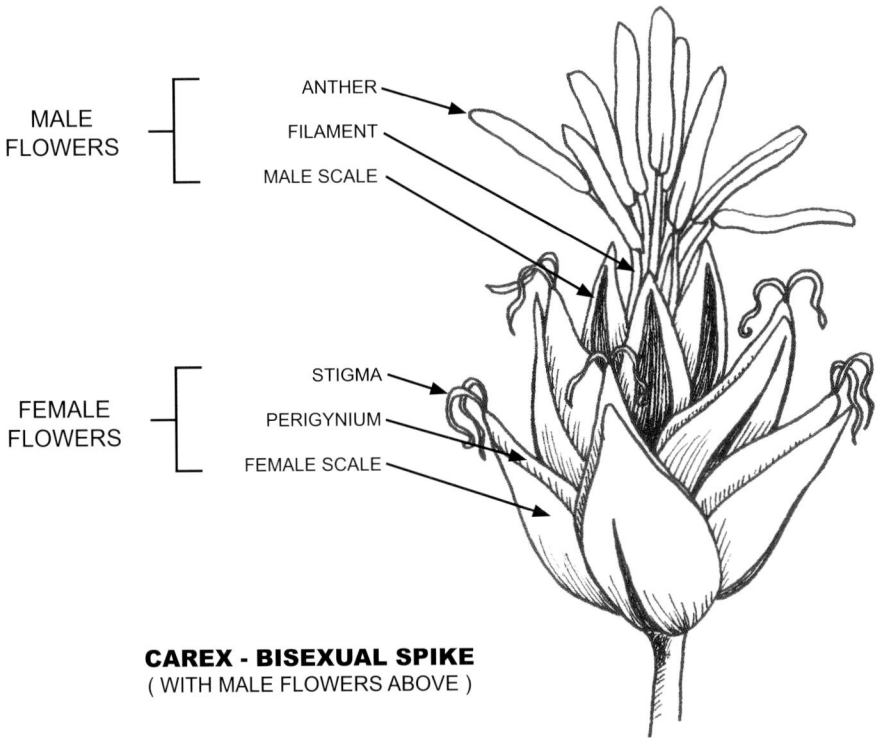

MALE FLOWERS

- ANTHER
- FILAMENT
- MALE SCALE

FEMALE FLOWERS

- STIGMA
- PERIGYNIUM
- FEMALE SCALE

CAREX - BISEXUAL SPIKE
(WITH MALE FLOWERS ABOVE)

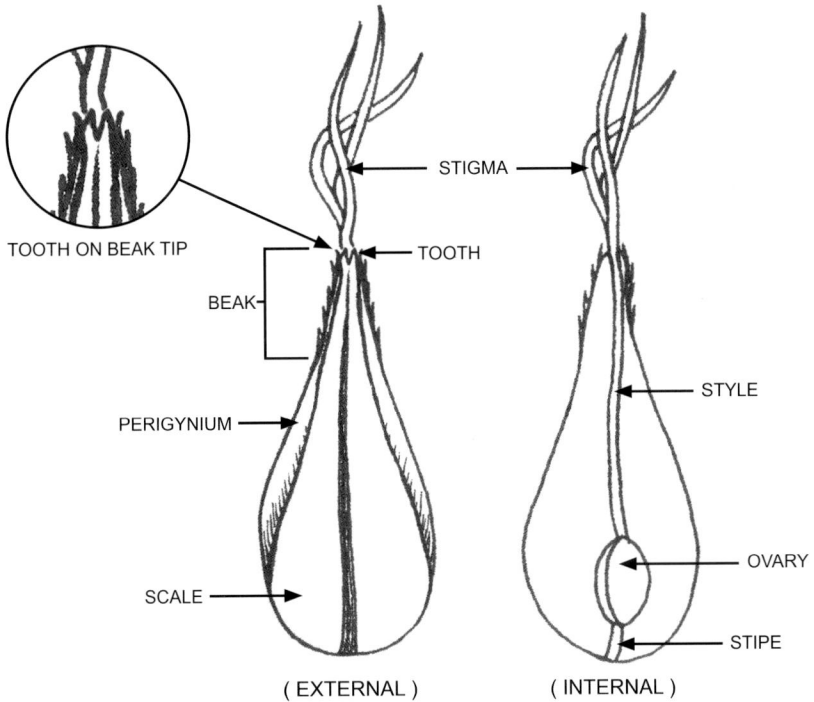

TOOTH ON BEAK TIP

STIGMA

TOOTH

BEAK

PERIGYNIUM

STYLE

SCALE

OVARY

STIPE

(EXTERNAL)

(INTERNAL)

CAREX - FEMALE FLOWER

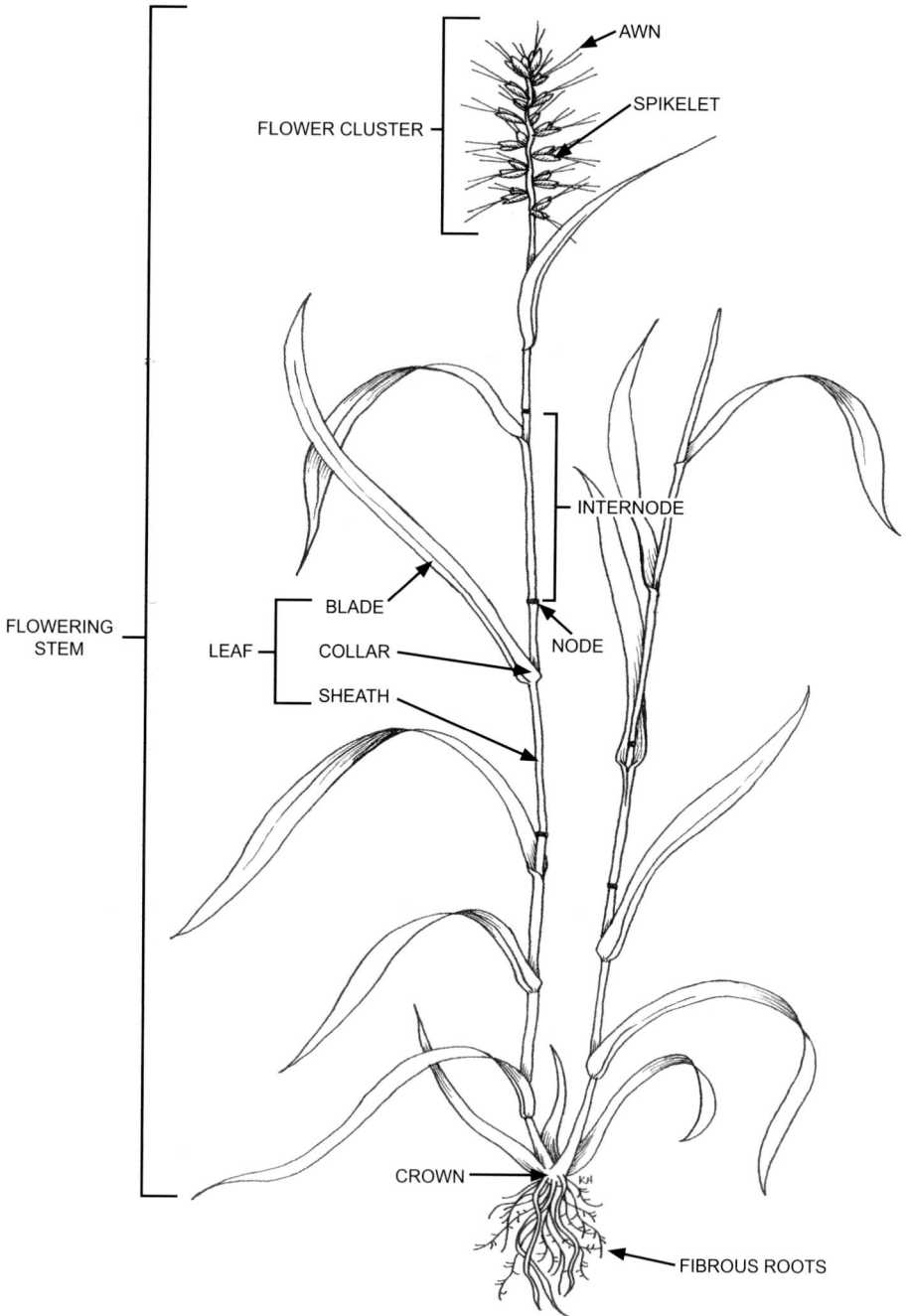

AWN

SPIKELET

FLOWER CLUSTER

INTERNODE

BLADE

LEAF

COLLAR

NODE

SHEATH

FLOWERING
STEM

CROWN

FIBROUS ROOTS

GRASS PLANT

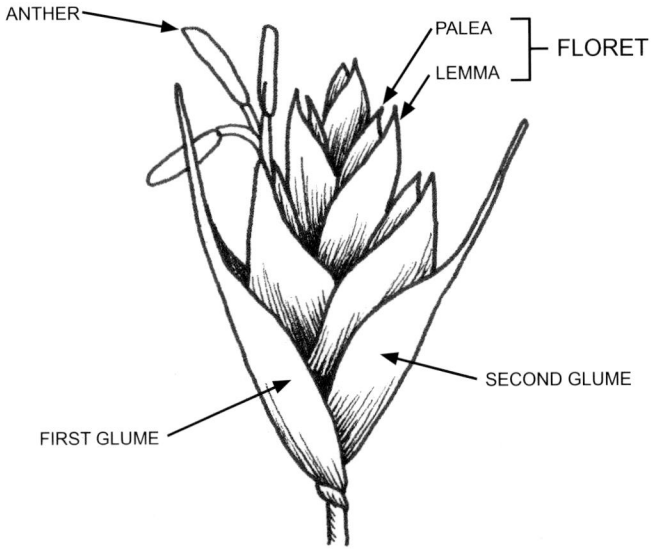

GRASS SPIKELET WITH FIVE FLORETS

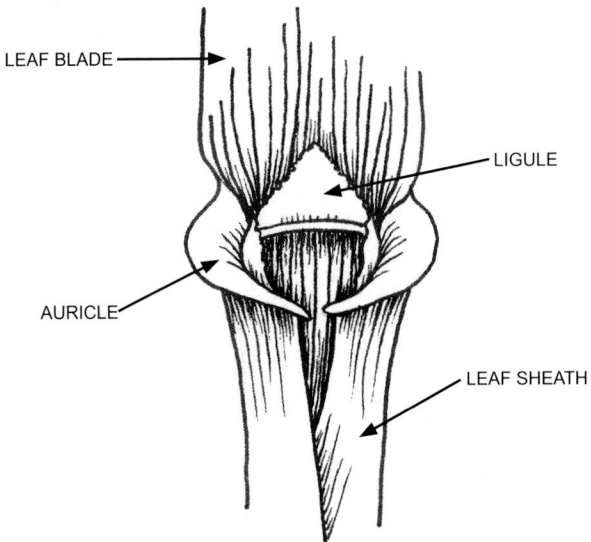

**GRASS - ENLARGED IMAGE OF JUNCTION
OF LEAF BLADE AND SHEATH**

REFERENCES

Barth, R. E. and N. S. Ratzlaff. 2004. *Field Guide to Wildflowers – Fontenelle Forest and Neale Woods.* Bellevue, Nebraska: Fontenelle Nature Association.

Evans, H.E. 1997. *The Natural History of the Long Expedition to the Rocky Mountains, 1819-1820.* New York; Oxford: Oxford University Press.

Garabrandt, G. W. 1976. *A History of Land Use in the Oak-Hickory Woodland of Fontenelle Forest. M.A. thesis, University of Nebraska at Omaha: 143p.*

Garabrandt, M. M. 1988. *An Annotated List of the Vascular Plants of Fontenelle Forest and Neale Woods in Eastern Nebraska.* Transactions of the Nebraska Academy of Sciences, XVI: 31-49.

Gilmore, M. R. 1919. *Uses of Plants by the Indians of the Missouri Valley Region.* Reprinted and enlarged Edition (1977). Lincoln, Nebraska: University of Nebraska Press.

Great Plains Flora Association. 1986. *Flora of the Great Plains.* Lawrence, Kansas: University Press of Kansas.

Haddock, M.J. 2005. *Wildflowers & Grasses of Kansas, a Field Guide.* Lawrence, Kansas: University Press of Kansas.

Hurd, E.G., N.L. Shaw, J. Mastroguiseppe, L.C. Smithman and S. Goodrich. 1998. *Field Guide to Intermountain Sedges.* General Technical Report RMRS-GTR-10. Ogden Utah: United States Department of Agriculture, Forest Service, Rocky Mountain Research Station.

Johnson, J.R. and G.E. Larson. 1999. *Grassland Plants of South Dakota and the Northern Great Plains.* Brookings, South Dakota: South Dakota State University, College of Agriculture and Biological Sciences, Ag. Communications.

Knobel, E.; Revised by M.E. Faust. 1977. *Field Guide to the Grasses, Sedges and Rushes of the United States.* New York: Dover Publications, Inc.

Kaul, R. B., D. M. Sutherland and S. B. Rolfsmeier in ms 2007. *The Flora of Nebraska.*

Kurz, D. 2003. *Trees of Missouri.* Jefferson City, Missouri: Missouri Department of Conservation.

Ladd, D. 1995. *Tallgrass Prairie Wildflowers: A Field Guide.* Guilford, Connecticut: Globe Pequot Press.

Maximilian Prinz zu Wied. *Reise in das Innere Nordamerika* (2 Volumes). 1841. Reprinted by Verlag Lothar Borowsky, Munich, Germany.

Moulton, G. E., Editor. 2002. *The Definitive Journals of Lewis and Clark.* Lincoln, Nebraska: University of Nebraska Press.

National Audubon Society *Field Guide to North American Trees, Eastern Region.* 1980. New York: Alfred A. Knopf.

Omaha Botany Club. 1959. *Plants of Fontenelle Forest.* Bellevue, Nebraska: Fontenelle Forest Association.

Pohl, R.W. 1978. *How to Know the Grasses, Third Edition.* Dubuque, Iowa: The Pictured Key Nature Series; Wm. C. Brown Company.

Rolfsmeier, S.B., 1995. *Keys and Distributional Maps for Nebraska Cyperaceae, Part 1: Bulbostylis, Cyperus, Dulichium, Eleocharis, Eriophorum, Fuirena, Lipocarpha, and Scirpus.* Transactions of the Nebraska Academy of Sciences, 22: 27-42.

Rolfsmeier, S.B. and B. Wilson, 1997. *Keys and Distributional Maps for Nebraska Cyperaceae, Part 2: Carex and Scleria.* Transactions of the Nebraska Academy of Sciences, 24: 5-26.

State of Nebraska Department of Agriculture, 1979. *Nebraska Weeds.* Lincoln, Nebraska.

Stubbendieck, J., S. L. Hatch and C. H. Butterfield. 1997. *North American Range Plants.* Lincoln, Nebraska: University of Nebraska Press.

Stubbendieck, J., M.J. Coffin and L.M. Langholt, 2003. *Weeds of the Great Plains.* Lincoln, Nebraska; Nebraska Department of Agriculture, Bureau of Plant Industry.

Yatskievych, G. 1999. *Steyermark's Flora of Missouri, Volume 1.* Missouri Department of Conservation, Jefferson City, Missouri in cooperation with the Missouri Botanical Garden Press, St. Louis, Missouri.

Index

honey, 54
Long-beaked sedge, 99
Lolium arundinaceum, 146
Lonicera
  dioica, 91
  maackii, 77
  tatarica, 76
Lovegrass, Carolina, 175

M

Maack honeysuckle, 77
Mace sedge, 124
Malus sp., 61
Mannagrass, fowl, 139
Marsh muhly, 173
Marsh oats, 182
Marshgrass, tall, 180
Maple
  silver, 2
  sugar, 60
Mead's sedge, 113
Menispermum canadense, 92
Mexican muhly, 186
Millet
  Chinese, 167
  green, 166
  wild, 166,168,177
Moonseed, 92
Morningstar sedge, 124
Morus
  alba, 34
  rubra, 32
Muhlenbergia
  bushii, 186
  frondosa, 186
  mexicana, 186
  racemosa, 173
  schreberi, 188
  sylvatica, 186
Muhly
  green, 173
  forest, 186
  marsh, 173
  Mexican, 186
  nodding, 186
  wirestem, 186
Mulberry
  red, 32

white, 34
Multiflora rose, 82

N

Nimblewill, 188
Nodding fescue, 138
Nodding foxtail, 167
Nodding muhly, 186
Northern catalpa, 38
Nutsedge, yellow, 129

O

Oak
  bur, 4
  red, 6
Oats
  blackbird, 182
  marsh, 182
Obovate beakgrain, 160
Ohio buckeye, 58
Olive
  autumn, 74
  Russian, 42
Orchard grass, 142
Oryzopsis racemosa, 159
Ostrya virginiana, 30

P

Pale bulrush, 125
Pale dogwood, 67
Panicgrass, Scribner's, 141
Panicum, fall, 193
Panicum
  capillare, 192
  dichotomiflorum, 193
  oligosanthes
    var. *scribnerianum,* 141
  virgatum, 179
Parthenocissus
  quinquefolia, 94
  vitacea, 95
Pascopyrum smithii, 150
Path rush, 130
Pawpaw, 37
Peachleaf willow, 26
Phalaris arundinacea, 143
Phleum pretense, 162
Phragmites australis, 165
Pigeongrass, 166,167,168

rubra, 16
thomasii, 18

V

Viburnum opulus, 84
Virginia creeper, 94
Virginia wild rye, 153
Virgin's bower, 96
Vitis riparia, 88

W

Wahoo, 75
Walnut, black, 43
Watergrass, 177
Wedgegrass, slender, 137
Western buckeye, 58
Western wheatgrass, 150
Western witchgrass, 193
Wheatgrass
   tall, 181
   western, 150
Whitegrass, 161
White ash, 46
White mulberry, 34
Wild black currant, 70
Wild currant, 70
Wild gooseberry, 71
Wild honeysuckle, 91
Wild millet, 166,168,177
Wild plum, 64
Wild rice, 182
Wild timothy, 173
Willow
   black, 28
   golden weeping, 60
   peachleaf, 26
Winged burning bush, 86
Wiregrass, 183
Wirestem muhly, 186
Witch's hair, 192
Witchgrass, 192
   western, 193
Wood grass, 160
Woodbank sedge, 117
Woodbine, 95
Woodland bluegrass, 135
Woodland sedge, 100
Woodreed, common, 176

Y

Yellow bristlegrass, 168
Yellow foxtail, 168
Yellow nutsedge, 129

Z

Zanthoxylum americanum, 63
Zigzag grass, 193
Zizania palustrus, 182